Emergency Medical Services (EMS) Recruitment and Retention Manual

 Federal Emergency Management Agency

 United States Fire Administration

Preface

The United States Fire Administration (USFA) publication Emergency *Medical Services (EMS) Recruitment and Retention Manual* is a guidebook intended for the use of the managers and recruiters of volunteer personnel in organizations that provide emergency medical services. The manual also may be used productively by EMS organizations. that employ career personnel, especially combined career-volunteer departments.

The manual was developed for the United States Fire Administration under FEMA contract EMW-93-C-4132 by the Transportation Research Center of the School of Public and Environmental Affairs at Indiana University in Bloomington, Indiana. The project development team and authors of the manual were James A. Palmer, Project Coordinator and Research Attorney, Transportation Research Center; John M. Buckman, Chief, German Township (Indiana) Volunteer Fire Department, Evansville, Indiana; and Lois R. Wise, Associate Professor, School of Public and Environmental Affairs.

Robert ("Red") McKeon, Chief, Occum (Connecticut) Volunteer Fire Department, and former Chairman of the National Volunteer Fire Council, and Jack W. Snook, Chief, Tualatin Valley (Oregon) Fire and Rescue Department, also contributed to this project.

State-level EMS departments and local-level EMS organizations in the United States were contacted by telephone by the project team during the course of the project in order to identify and learn about innovative, interesting, and effective strategies being used to recruit and retain EMS volunteers. The manual could not have been completed without the information and insights about EMS recruitment and retention that were contributed by the contact persons with these agencies.

A Quality Review Panel was established to review and comment on the first and second drafts of the manual. The members of the Quality Review Panel were Mary Beth Michos, Chief, Prince William County (Virginia) Department of Fire & Rescue, Rohn M. Brown, Recruitment and Retention Coordinator, Office of Emergency Medical Services, Virginia Department of Health; and Ted Halpin, Executive Director, Ontario County (New York) ALS/Thompson Health System.

Contents

requirements were established to improve the performance, image, and attitude of prehospital EMS personnel. The training requirements set by government regulators dramatically increased the time demands on volunteers, and volunteer EMS agencies not used to medical oversight began to feel the pressure from physicians and hospitals to improve the quality of "street medicine."

Physician and hospital involvement in improving the quality of emergency medical services created friction in the relationship between volunteer EMS agencies and the medical community. In addition, government regulations concerning EMS operations, personnel, and equipment were seen as an unnecessary outside interference with the right of volunteer EMS agencies to manage and control their organizations. Volunteers have a great sense of pride, and some initially resented this intervention by the medical community and government regulators. However, as EMS organizations upgraded their training and capabilities and became more accustomed to medical and governmental oversight, the volunteer resentment diminished and cooperative relationships with the medical community and government regulators were created.

American society is changing rapidly, and some of these societal changes directly influence volunteering in the United States. In addition to these nationwide influences, there are local influences unique to each volunteer EMS organization: the community in which the service is provided, the organization itself, and the individuals involved. Each of these influences--societal, community, organizational, and individual--must be addressed when you answer the question: "What factors influence our volunteer program?"

Volunteerism in the United States

Volunteers are characterized by the diversity of their backgrounds, ages, gender, and reasons for volunteering. Since almost one-half of the population engages in volunteer activities, it is no surprise that the "typical" volunteer can be almost anyone.

Volunteers are of all ages, with the largest number between 30 and 45. Almost a quarter of the population under age 30 and over age 65 is involved in volunteer work. Each sex is represented equally among volunteers; almost one-half of all males and females are involved in volunteer activities. Both white-collar and blue-collar workers, as well as students and retirees, are represented among volunteers.

Individuals become involved as volunteers for a variety of reasons. The most frequently cited reasons for volunteering are:

- ' "Wanted to help others."

- ' "Felt obligated to give back what I got."

- ' "Sense of citizenship."

- ' "Religious feelings."

- ' "Interested in the work being done."

- ' "Desire for involvement with a group of friends."

Volunteers usually experience a variety of motivations, ranging from altruism to self-interest, during their volunteering "career," and their motivation indeed may vary considerably as they work over time within a single organization.

The Role of Leadership in Volunteer Programs

The key to success in recruiting and retaining volunteers is effective and dynamic leadership. Programs involving volunteers do not work spontaneously. They require a leader to provide the care and attention needed to fit together a complex system for delivering services that matches the needs of the community and the program. The role of the leader includes working with the program's staff, funding sources, and clientele to identify what needs to be accomplished, which activities can be performed by part-time volunteer workers, and what will motivate those volunteers to participate. The effective performance of the leadership role requires someone who relates well to people and understands both their needs and wants, is flexible and can adapt to changing demands and interests, and is able to identify potential opportunities and to create the environment necessary to take advantage of those opportunities.

The management of volunteer programs is a growing professional field. While many volunteer programs are managed effectively by volunteer leaders, many organizations now pay staff to assume responsibility for their volunteer programs. Program directors who work with volunteers are hired by governments to manage many EMS delivery systems. Much time and energy is involved in meeting both program and volunteer needs. When a volunteer fills the primary management role in an organization, there may be a problem with continuity in leadership. A full-time paid director provides the needed continuity in management and is in a strong position to represent the interests of volunteers with both paid personnel and the general community. However, even a full-time paid position may not be sufficient to meet all of the needs of a sizable EMS system.

EMS organizations must compete with other community organizations to attract the most capable and committed volunteers. The organizations that are well managed will be the most appealing to prospective volunteers. Volunteers are more likely to join and remain with well-run organizations with competent leaders.

Planning for the Participation of Volunteers	Volunteer involvement has a unique place in our history. Helping one another is an American tradition. Volunteering in EMS organizations today has changed because our society has changed, but the basic commitment to helping one another remains the same. EMS leaders must be aware of and respond to these changing conditions and trends in the future.

The first step in determining an organization's need for volunteers is to develop a plan with two questions: Why does the organization need volunteers? How many volunteers are needed? The answers to these questions will enable the organization to:

- Determine the specific jobs and tasks that volunteers will perform.

- Explain to volunteers how and why they are contributing to the mission of the organization.

- Explain to any career staff why volunteers are being sought.

- Develop a plan to evaluate whether the use of volunteers has been effective.

Why have volunteers? If an organization consists only of volunteers, the answer is obvious. Without volunteers the organization would not exist. In organizations that could function without a volunteer program, volunteers can still contribute significantly to the accomplishment of the organization's goals.

In the planning process, it is important to list the positive and negative points of volunteer participation in the organization. Both the positives and negatives need to be acknowledged. The positive factors make it worthwhile to spend energy on effective management. Some negative factors can be altered by sound planning and leadership, but others cannot and must be viewed realistically. If the positive points of volunteer involvement do not outweigh the negative points, effective leadership and management may be insufficient to compensate for the disadvantages, and perhaps volunteers should not be involved in the organization.

If everyone in the organization is committed and excited about having volunteers, it is not necessary to spend much time in enlisting internal support for volunteer participation. If there are complaints about having volunteers in the organization, an attempt must be made to enlist the support of all members of the organization to mediate these concerns.

If volunteers are going to work with career staff, it is essential that staff are in agreement as to the purpose and worth of the volunteer program. It is important for all levels of career and volunteer staff to be involved in the planning process.

It is also desirable to have the support of top management of an organization for the volunteer program. This support can be demonstrated by the official adoption by the governing body of the organization of a policy endorsing the use of volunteers or by a position statement on volunteers approved by the director of the organization.

Recruitment

Recruitment is a key to the success and survival of an EMS organization. It is the systematic, continuous process by which capable volunteers are identified, evaluated, and enlisted as members of the organization. The major considerations involved in the development or improvement of an effective EMS recruitment program are presented in this unit of the manual: the multi-step process for recruiting and selecting volunteers! the use of a dedicated recruiter, the relevant federal laws affecting recruitment, and the need for and benefits of diversity in an EMS organization. The next unit--*Motivation*--also provides information concerning recruitment: how to attract and persuade individuals to volunteer as members of an EMS organization. Specific techniques for recruiting, as well as retaining, volunteers are described in the unit *Recruitment and Retention Strategies.*

Steps in Recruiting EMS Volunteers

The process for the recruitment of volunteers for an EMS organization can be viewed as a series of steps. By following these steps, an EMS organization will have an efficient system for recruiting new volunteers, keeping everyone informed about the status of the recruitment effort, and evaluating the strengths and weaknesses of applicants for volunteer positions.

An important consideration in the recruitment process is the periodic assessment of the organization's needs for volunteers. A needs-assessment strategy involves thinking about present needs, as well as future needs that are likely to develop, and planning for how those needs will be met: What kinds of tasks do you need volunteers to perform? What capabilities or skills are required? Are there sufficient volunteers with the needed capabilities to perform those tasks? What additional training is required? How many additional volunteers are required to undertake those tasks now and in the future?

If a volunteer member is planning to move or take a job that precludes or reduces his or her participation, the organization may have a present need to replace the member by recruiting another volunteer or by retraining or reassigning an existing volunteer. If the organization has a general idea based on experience as to how long, on the average, volunteers serve, it has identified a future need resulting from turnover that can be anticipated, and the recruitment program can be designed to ensure that an adequate number of new volunteers are enlisted in the future. Strategic thinking today about how the work can be organized to get the most out of available volunteers and how upcoming needs can be addressed will enable the manager to operate the EMS organization more efficiently and effectively.

The basic steps to follow for recruiting and selecting volunteers are:

- Develop and implement a needs assessment based on the EMS organization's current volunteer staffing, existing vacancies, and anticipated need for staffing, including daytime volunteers.

- Identify the skills, knowledge, and abilities needed and any specific certifications required.

- Prepare job descriptions based on tasks and responsibilities.

- Develop a plan and timetable for the recruitment of the various types of volunteer personnel and skills needed.

- Implement the recruitment plan for the volunteer personnel currently needed. Provide feedback, encouragement, and information to potential volunteers during the recruitment process.

- Implement a system for evaluating potential volunteers that is compatible with applicable civil rights laws. The evaluation system used may be an existing system or a new system developed by the organization; however, the procedures employed in the system must be valid and reliable. Skill or knowledge tests, if appropriate, or open-ended interview forms should be administered. Tests based on the performance of real tasks offer the most reliable information about the relative abilities of candidates for volunteer positions and are more easily shown to be valid if challenged in court.

- Rate or rank order the applicants based on established criteria.

- Select the candidates with the best qualifications.

- Develop a schedule compatible with current and anticipated needs for volunteers for bringing the selected candidates on board. Schedule followup contacts with qualified candidates placed on a waiting list.

- Implement an orientation and training program for new volunteers.

- Assess the progress of new volunteers and make recommendations as to changes needed in performance or training.

Use of a Dedicated Recruiter	The responsibility for an EMS organization's recruitment efforts should be assigned to a dedicated recruiter or volunteer coordinator, who does not need to be an operational member of the organization, in order to ensure a continuous, professional approach to the identification and enlistment of a competent volunteer workforce for the organization.

The recruiter selected must be enthusiastic, dedicated, motivated, and sold on the product, which is the importance of the EMS organization and its work to the community. A casual attitude towards recruitment will not achieve the results needed. It is more important for the EMS organization to take the time to select the right person to sustain the recruitment program with the right attitude than simply to assign the job of coordinating the recruitment effort to anyone in the organization.

The dedicated recruiter is responsible for developing a program to recruit individuals into the organization. The recruiter is not solely responsible for the actual recruiting. Recruiting is the job of every member of the EMS organization. All members must participate in implementing the recruitment program and in identifying and attracting capable individuals who express an interest in the organization.

One of the first things that a recruiter should do is to train other members of the organization how to recognize an opportunity to recruit and to follow through with recruitment opportunities. The recruiter should develop print materials that explain the organization and its mission and activities and ensure that these materials are available for distribution by the membership and in public facilities throughout the community.

The assignment of a dedicated recruiter is vital to the overall success of the EMS organization. The person assigned this task is as important to the organization as other staff members. If proper attention to the manpower needs of the EMS organization is not maintained, EMS managers can expect problems in maintaining adequate staffing levels and the organization's ability to perform its mission, and the morale of its membership will suffer.

**Some Relevant
Laws Affecting
EMS Recruitment**

Depending upon the state in which an EMS organization operates, different laws pertaining to the conditions of employment may apply when an EMS manager makes decisions about recruiting, selecting, and using volunteer EMS personnel. It is not feasible to list the relevant laws of all fifty states, but it is possible to provide a brief summary of the most important federal laws affecting employment.

Fair Labor Standards Act

The Fair Labor Standards Act, which has been implemented by the Department of Labor (DOL) since 1938, is designed to protect employees from being coerced by their employers into volunteering at their place of employment. The FLSA was extended through amendments in 1966 and 1974 to include most state and local government employees. Fire departments and EMS organizations need to understand that FLSA is the law and DOL is actively enforcing the FLSA regulations. Noncompliance continues to cost fire and EMS departments money both in violation fines and in compliance costs, which are usually retroactive.

The purpose of FLSA is to set minimum wage and overtime pay standards for employees. These requirements are enforced by DOL's Wage and Hour Division (WHD), and its compliance efforts center around employee complaints of employer's alleged violations of FLSA. DOL can initiate legal action against employers who violate FLSA to recover back wages due employees and can also seek to restrain employers from future violations.

In the 1985 case *Garcia v. San Antonio Metropolitan Transit Authority*, 469 U.S. 528, the U.S. Supreme Court held that the Congress does have the authority to apply the FLSA minimum wage and overtime pay requirements to state and local government employees who are engaged in traditional governmental activities. As a result, state and local units of government are required, among other things, to compensate those government employees in traditional activities who work overtime with cash wages rather than compensatory time off,

Volunteers who receive no compensation and perform wholly charitable services, such as mentoring a needy child, are not subject to the FLSA requirements. Volunteers who receive some type of compensation, whether monetary or in-kind, may be entitled to additional pay or overtime consideration. FLSA applies only to individuals considered to be "employees. " The act itself offers little guidance for determining whether someone is an employee or a volunteer. There is no definition for the term "volunteer." An "employee" is "any individual employed by the employer"; "employ" means "to suffer or permit to work. "

To determine whether an employment relationship exists for the purposes of the FLSA, DOL looks at the following four major factors: education, benefits, competition for open positions, and wages.

Each EMS organization is encouraged to contact its local legal counsel for advice as to its potential liability under FLSA.

Title VII of the Civil Rights Act of 1964

Title VII of the Civil Rights Act of 1964 provides the basis for equal employment opportunity and forbids discrimination in employment on the basis of race, color, national origin, religion, and sex. By interpreting Title VII broadly, the courts have set precedents affecting a number of recruitment and retention issues, especially in the area of testing. Because they are used as a basis for making choices, interviews are considered a test instrument under this law and must be free of discrimination.

Selection tests for new recruits and for placement according to law must be job-related. If members of certain groups are more likely to be turned down than others, the fairness of the selection procedure should be reexamined.

Equal Pay Act of 1963

The Equal Pay Act of 1963, which is an amendment to the Fair Labor Standards Act, prohibits different rates of pay for equal work based on a person's sex. This law applies when:

- the working conditions are the same;
- the jobs are the same; and
- the jobs require similar skills, responsibilities, and efforts.

Differences in pay between a male and female EMT doing the same work could be justified, for example, if they are the result of any of the following:

- a seniority system;
- a merit system;
- a production-based pay system; or
- a factor other than sex.

Age Discrimination in Employment Act (ADEA) of 1974

The Age Discrimination in Employment Act protects workers who are forty years of age or older from discrimination in applying for employment or from discrimination on the job. It is a violation of federal law to discharge individuals at a specific chronological age unless a mandatory retirement age exists for their occupational category. Assumptions may not be made about an individual's physical abilities based on his or her chronological age. In fact, preselection queries regarding age, date of birth, or other proxies are a violation of ADEA.

Americans with Disabilities Act of 1992

The Americans with Disabilities Act of 1992 (ADA) extended to individuals with disabilities the civil rights protections that are afforded to others under Title VII of the Civil Rights Act. As in the case of the Title VII protections for minority group members and women, the full implications of the act will depend on court interpretations of the law.

There are two important aspects of the ADA that affect EMS organizations. The first is the requirement that many public facilities must be accessible to individuals with disabilities. This mandate may have consequences for facilities of an EMS organization; it may mean that it will be easier to transport individuals using wheelchairs and other equipment.

The law also forbids employers from discriminating against individuals with disabilities who can perform the essential tasks of a specific job, even if they cannot perform other marginal job duties, if an employer can make a reasonable accommodation. The ADA represents a very important change in policy towards individuals with disabilities even though its application to volunteer workers is still unclear. The fact that individuals with disabilities have valuable skills and talents to contribute to an EMS organization is the key point for managers of volunteer services to consider,

Diversity in EMS Organizations

In many communities in America, it is apparent that the population is becoming more diverse. It will be difficult and unwise to carry out a recruitment program based on the traditional image of an EMT volunteer; there just aren't enough of them. EMS managers should value diversity by recognizing that capable volunteers come in different packages with different personal lifestyles, values, and priorities, by reaching out to the entire population of potential volunteers when recruiting new members, and by not making assumptions about whether or not individuals would or could perform EMS tasks based upon their age, sex, race, ethnicity, or disabilities.

Older individuals will become increasingly important as volunteers as our population continues to retire earlier and enjoy good health. Women are often under-represented in EMS organizations even though they are as competent as men in serving as EMS volunteers. Members of different racial and ethnic groups want to participate in the network of active citizens just like everyone else, but need extra encouragement to become involved. Individuals with disabilities can perform many tasks that are of great value to EMS organizations.

A strategy for recruitment that reflects the philosophy of valuing diversity means developing and executing a plan to reach out to nontraditional volunteers. It means giving everyone equal access to membership, training opportunities, and decisionmaking and listening to their ideas, solutions, and concerns. The participation of nontraditional volunteers can both enrich and improve the operation of an EMS organization.

Valuing diversity also means taking responsibility to ensure that individuals from different backgrounds are integrated into the organization and made to feel they are an important part of the EMS team. The EMS manager should be aware that these individuals may be sensitive about being different or not one of the group and that adjustments may need to be made to accommodate them. For example, the organization's facilities may need to be modified so they are wheelchair-accessible, and special equipment may be required to enable individuals with disabilities to field phone calls. The range of sizes for uniforms, boots, or other equipment may need to be increased.

Diversity in cultural backgrounds raises some other issues about managing volunteers. Individuals from some cultures may not react to praise or criticism the same way as others or may be reluctant to claim responsibility for their own accomplishments. Religious holidays or family priorities may need to be taken into consideration when scheduling certain volunteers for work. EMS managers will need to be aware of such cultural differences when evaluating, motivating, and encouraging volunteers. Flexibility is required in managing a diverse volunteer membership.

The creation of an environment that welcomes cultural diversity may require some special effort. Special training concerning issues related to diversity may need to be provided for the membership. Trainers with expertise in this field are available to provide special diversity programs, films, and group problem-solving exercises. There is a risk, however, that diversity training can be counterproductive by bringing to the surface but not resolving personal prejudices.

Creating the Right Environment

Valuing diversity means making individuals feel welcome. Jokes about one's heritage or appearance may not be appreciated, and expressions used without any thought about their content may be offensive. Some signs, calendars, and wall art may be perceived as degrading or embarrassing. Such organizational behavior will put some members on the defensive or make them feel inferior. An EMS organization that tolerates racial or ethnic bigotry, personal slurs, or sexual harassment is not a welcome place.

There are no secret formulas for changing the culture and attitudes of an organization, but one thing is critical to any change: **EMS managers must lead by example.** A manager cannot practice or tolerate behavior that is offensive to minority groups, females, seniors, or disabled volunteers. A manager cannot change people, but can demonstrate what is allowable in the EMS organization.

Motivation

The purpose of recruitment is to motivate individuals to volunteer their capabilities, time, and energy for the benefit of the EMS organization and the accomplishment of its lifesaving mission. EMS managers and recruiters need to understand the basic principles and techniques for marketing their organization and motivating potential recruits to enlist as volunteers.

Motivation is a comprehensive topic. There are many different theories about HOW to motivate people to join an organization and even more theories about WHAT motivates people. Let's start with HOW.

Motivating Them to Volunteer

Most people really do things based on impressions they have rather than through some thoughtful analysis of the pro's and con's of a course of action. Decisions about joining voluntary organizations often come about that way. People get an impression that it would be challenging or interesting to serve as an emergency medical technician, and something confirms that idea and motivates them into action.

How do people get impressions that make them think about joining an EMS organization? Positive impressions can be gained through relatives, friends, or coworkers who are active volunteers, through public-relations advertising, and through television programs or movies about EMS. In addition, EMS managers can create situations that make a positive impression through special events and promotions. A variety of approaches for persuading individuals to volunteer may be required because, according to the "three times rule," a message must be heard or seen at least three times before an individual will be persuaded to act.

Once this idea has been planted in someone's mind, he or she can be moved to action by a friend, relative, or co-worker who actively recruits them, by an event the EMS organization creates in which people come forward to explore the possibility of volunteering, and by mailings, phone calls, or advertisements designed to persuade people to enlist.

Some individuals are more rational. They volunteer for a specific reason, like getting experience that will help them in their career. These individuals will think more about the benefits of being a member of an EMS organization and may want more information describing just what these advantages are. They may want to know what kind of equipment they will become qualified to use and what sort of emergency medical training or certification they will receive from their volunteer work. They may want to know whether anyone before them has had success in converting his or her volunteer experience into regular employment. They also may want to know more about the disadvantages or costs to them, such as the equipment they will have to buy, the time they will need to invest in training, or the time and expense required for recertification in the future. They usually want enough information to make a reasonably informed decision. Booklets and information sheets can be very helpful here.

If the EMS organization is located where there is more than one volunteer service, some individuals may try to compare the advantages and disadvantages in order to decide which one offers them the best opportunity, For this reason, an EMS manager might want to keep abreast of the recruitment incentives that similar volunteer services are offering so the EMS organization can try to be competitive with them in attracting new people.

Word-of-Mouth Recruitment

EMS volunteers and their friends, relatives, and neighbors are excellent recruiters, especially when they have a positive EMS experience to share. They are the best recruiters not only because of their contacts with other people, but also because they can convey something about the nature of EMS work. This type of recruitment can be enhanced by providing volunteer members with professional business cards and basic organizational flyers. Volunteers recruited this way usually have a more realistic picture of what EMS work entails and even a familiar face in the organization to show them the ropes. As a result, they are less likely to drop out in the initial period after their enlistment.

One down side to relying on existing EMS volunteers and their friends, neighbors, and relatives is that it tends to limit the cultural diversity of the volunteer force because many individuals limit their social contacts to others like themselves. If an EMS organization relies on social networks for volunteers, it probably will want to make a special effort to reach out to under-represented groups in the community for volunteers as well.

Getting Their Attention: Marketing

Marketing has been around since time began, but it's been going by other names, such as sharing, trading, exchanging, or bartering. It is the caring trade of value for value. For an EMS organization that needs volunteers, marketing involves the identification and offer of something valued by volunteers in exchange for their uncompensated services.

The true worth of marketing comes when all parties involved in an exchange relationship are convinced they have received the greatest value. People can relate to value in many instances only when they literally can feel the value. In emergency medical services this may be hard to do, but is not impossible.

Value can be felt by the community in several ways. One way is by receiving emergency care or other services from the EMS organization; another way is by learning about an EMS organization's activities through an effective public-relations program.

Like any of the tools needed to manage a well-run program, marketing requires hard work, creative problem-solving, and innovative thinking to be successful.

The Importance of Marketing

Marketing is important to the overall success of an EMS organization:

1. *In recruitment.* Marketing is the bridge between potential volunteers and the EMS organization. It defines what is needed and what benefits are available, in addition to the incentives offered to prospective volunteers, to attract and involve them in the organization.

2. *In volunteer retention.* The value given to volunteers in tangible and intangible rewards is what keep volunteers coming back and stimulates them to tell others about their good feelings, which feeds back into further recruitment.

3. *For the organizational climate.* Happy volunteers who feel good about their work create a climate that stimulates creativity, enjoyment, and achievement. It simply feels good to work and be there.

4. *In gaining organizational support.* When an EMS manager needs the support of the organization's administration, board members, peer-level department heads, and volunteers, marketing provides the basic leverage that allows the manager to define an exchange relationship (what's in it for them and for the manager) to attain the manager's goals.

5. *In winning public support.* EMS managers need the support of the general public for the organization's programs, events, and

philosophies. Marketing again holds the key to attaining this support. Public relations can play a key role here.

6. *In resource development.* When an EMS manager needs goods, services, or dollars, the marketing orientation is the magic that will produce results. In fact, marketing is the key to getting what the manager needs without spending dollars: the art of fundraising.

7. *In obtaining members, participants, and consumers.* Marketing is also the key in getting people to accept the EMS organization's services and/or products. Proper application of marketing convinces others to become involved with the organization when they see there is great value to themselves for that involvement.

Too often, marketing is the missing tool in the EMS manager's toolbox, with the result that issues begin to arise concerning recruitment, retention, and fundraising. As shrinking resources, increased requirements for volunteers' time, and greater demands from the public affect EMS organizations more dramatically each year, EMS managers must master this tool in order to achieve the EMS organization's goals.

The Three Components of Marketing

Marketing is made up of three components: publics, markets, and the exchange relationship. Let's define all three.

1. *Public.* A public is any identifiable segment of the community in which the EMS organization operates. Publics are groups with whom the EMS organization might or might not want to have a relationship. When identifying publics, the EMS manager doesn't worry about or try to judge whether or not he or she might someday want to interact with them. They simply exist.

2. *Market.* A market is an identified public with whom the EMS organization decides it wishes to establish a trade relationship. To put it in simpler terms: they have what the organization needs or wants. On a list of 100 publics, an EMS manager may find six markets that could meet the organization's needs if they choose to do so. The trick is having a list of publics from which to choose. Without it, EMS organizations suffer from resource nearsightedness, never realizing how many potential resources are available to them.

3. *Exchange Relationship.* This is the keystone of success. It is the bargain that is struck between the EMS organization and the markets. It is the essence of marketing: the trade of value for value. The characteristics of the exchange relationship are:

- Honesty and fairness.
- No hidden agendas or pitfalls.
- A user-oriented position.
- A targeted approach.
- Highest concern for what the other party will receive of value.
- Attention to the organization's success in attaining its goals.
- A lot of homework.

Planning a Marketing Program

When thinking about marketing the EMS organization, it is essential that the EMS manager understand the four-step planning process that will bring success.

> **The Four Steps in Planning a Marketing Program**
>
> Step #1: What do you have?
>
> Step #2: What do you need?
>
> Step #3: Who has what you need?
>
> Step #4: How do you get what you need?

In most instances, EMS managers begin with step #2--their needs; as a result, they find themselves with many problems, including resource nearsightedness. The manager must first think of what the organization has:

- What business are you in?

- What's your product?

- Three product classifications:

    - The expected product? It is critical to learn each volunteer's expectations before beginning work in order to compare them to reality.

    - The augmented product?

    - The potential product?

- What business does the public think you're in?

- What products does the public think you offer?

Four Ways to Market

There are many things that go into marketing, but there are four basic ways to market: advertising, publicity, promotion, and personal selling.

1. *Advertising.* Advertising is a campaign that is designed, pretested on target markets, posttested for results, and presented through selected media in order to influence people to accept a product or idea. It usually has a theme or slogan, is targeted to specific audiences, and has a cost. Too frequently, EMS organizations misunderstand the purpose of advertising, and consequently they omit testing, targeting, and life-cycle checks, thereby making their ads ineffective.

Advertising can serve an EMS organization by making people aware of what it does, what it needs, and how it can be contacted. Advertising usually is designed for quick consumption, not lengthy reading, and for creating familiarity.

Many EMS managers confuse advertising with marketing, thinking they have a wonderful recruitment campaign underway, for example, with 24 ads in newspapers and 100 posters in stores. This **is not a recruitment campaign;** it is an advertising effort that can augment recruitment through personalized asking, but by itself is not going to be successful in attracting volunteers or donors.

The most remembered and successful advertising of this century was the "Uncle Sam Wants You!" campaign that appeared in every post office and train station across the United States. Had the posters and ads, however, not been backed up by recruitment offices, appeals, the draft, and the fervor of people wanting to help their country, the ads themselves would have fallen short of their goal.

2. *Publicity.* Publicity is a cost-effective technique that involves the development and dissemination of news and promotional material designed to bring favorable attention to a product, person, organization, place, or cause. It is different from advertising in that it is usually carried by the media for free and without indication of the source; it also can be more credible than advertising. An active public information officer for the EMS organization is required to ensure maximum publicity.

Publicity is a creative challenge for an EMS organization as it finds ways to use media to tell its story and express its need for support. It is critical for an organization to check the effectiveness of its publicity by finding out how well the public is getting its message.

It is publicity that plays a key role in creating impressions and perceptions and needs to be current, clear, and realistic as it comes across to the public.

3. *Promotion.* Promotion is defined as marketing activities that project the EMS organization's message other than by advertising, publicity, and personal selling. Promotional efforts might include displays, speakers bureau presentations, and booths at local fairs and events. These efforts are usually unique and are not necessarily expected to occur again. The creative minds within the EMS organization can be stimulated to think of positive ways to gain attention and share information with the public.

4. *Personal Selling.* The effective EMS manager understands that some of the best recruiters for the organization are satisfied volunteers who relate their positive experiences and encourage others to become involved. Satisfied donors and supporters can and do play the same role through their contributions and their influence with others.

The catch in understanding personal selling, however, comes when this activity is separated into its two components: spontaneous and targeted. For the most part, the contacts made by volunteers and supporters are sporadic and unplanned. This is spontaneous personal selling and is difficult to measure or control. Most of the individuals involved in this form of selling do not even realize what it is; they are simply sharing their pleasure with others. The second type of personal selling--targeted selling--is planned and involves people who are interested in participating in events the public is expected to attend. The volunteers involved in targeted selling could even be coached as to what to say, how to act, and what to wear.

Communications Strategies

There are many ways to communicate the EMS organization's message. Depending on the demographics of the market or the audience, some communications strategies will work better than others.

General Considerations

It has been demonstrated that spending a lot of money doesn't necessarily pay off with big results. Small items listed in newspaper want ads may produce better results than custom-made ads on news pages. Similarly, an ad in a flyer handed directly to an individual may be more effective than the same ad run in a local newspaper. If there is more than one EMS organization in your area, you may want to pool resources and share the cost of advertising in print media. Some local newspapers periodically publish a listing of volunteer opportunities.

Direct ads that ask individuals to volunteer are one form of advertisement, but the media also can be used to give information about an EMS program and its activities that has a positive effect on recruitment. News coverage, photographs of accomplishments, and news stories that describe the role of EMS volunteers are important opportunities for conveying information that make a positive impression. For this reason, an EMS organization's manager or public information officer should maintain good relations with press reporters even though it is not always easy to respond immediately to their requests for information.

There are other ways to get the volunteer message out. If you have volunteers who like to speak to the public, their appearances before community groups are also excellent opportunities for recruitment. A listing of opportunities for volunteering may be kept by the public library or by the city or county government volunteer office. Local radio and TV stations may air public service announcements at little or no charge.

Message Content

The content of the message is as important as the way it is communicated. Short, memorable one-liners can capture a listener's or reader's interest and have the positive effect desired. The use of a visual image in combination with a one-line message is a very effective approach. An ad that enables someone to put himself or herself in the picture will attract those who are motivated by civic responsibility or social duty.

> **Sample Messages**
> - "You can make a difference"
> - "Volunteer for life"

Organizational Image

The facilities, vehicles, equipment, and uniforms of an EMS organization may capture the interest of some individuals and operate as an incentive for volunteering. Pagers, beepers, and mobile phones may make the opportunity to join an EMS organization appear more attractive and interesting to some volunteers. For this reason, the EMS manager should be conscious of the different kinds of equipment used, the cost of the equipment to volunteers, and the use of other technical apparatus that increases the job challenge. Potential volunteers may be reluctant to join an EMS organization if they have to pay for their uniforms and equipment.

Use the EMS PIER Manual

The *EMS PIER Manual* (Publication FA-151/September 1994) distributed by the U.S. Fire Administration and the National Highway Traffic Safety Administration provides guidance for EMS organizations in creating, maintaining, and enhancing their public information, education, and relations programs. This manual is an excellent, practical guidebook and is highly recommended for use by all EMS organizations. It may be obtained for free from either the USFA or NHTSA. Consult the *Resources* unit of this manual for the mailing address and telephone number for these federal agencies.

Recruitment Reminder: Be Realistic

Being realistic with potential recruits is especially important when they are volunteers, who usually have fewer and weaker ties to an organization than regular employees. When volunteers, especially EMS volunteers, decide the job is more than they bargained for, they can and will quit. Volunteers must not be given the impression that the job is all glamour and happy endings. They need to be aware of the difficult problems they will face so that when a crisis occurs they are not completely surprised. If the EMS organization fails to provide volunteers with sufficient notice as to the difficult situations EMTs may face, the unanticipated emotional trauma of an emergency run with an unsatisfactory result may lead to the volunteer's subsequent failure to show up for calls or even to resignation.

By the same token, potential volunteers should be encouraged to be honest about what they have and can do and what their own personal limitations are. Volunteers may develop an ability to deal with difficult situations, but if they are faced with tough problems they are not ready for, they are likely to quit. Volunteers may not know how they will react to the misfortune of others and should be encouraged to think this through and make a realistic assessment of what they can handle.

There are many ways to bring reality into the decision to volunteer. Shared experiences, films, and roleplaying exercises are techniques that can be used to develop a fair assessment by both the recruiter and a potential volunteer as to whether or not that individual should volunteer for EMS work. Realistic recruitment doesn't mean dwelling on the negative parts of the job, but it does mean providing an accurate picture of what is likely to happen. The bottom line is that volunteers themselves must have enough information about what they will do in the EMS organization to make a decision that is right for them.

Written job descriptions and signed commitment statements are useful tools for ensuring that both the volunteer and the EMS organization have shared expectations as to the volunteer's role in organizational activities.

What Motivates Them to Volunteer for EMS

People do different things for different reasons, and their reasons may change over time. There is no specific formula about what or how much of something makes a person decide to volunteer for EMS. Everyone is different. Nevertheless, many things are known about what motivates individuals to join and stay in voluntary organizations.

Opportunity for Friendships and Cooperative Activities

The opportunity to develop friendships and to work cooperatively with other people in a rewarding activity is a major reason why volunteers enlist and remain in EMS organizations. Therefore, it's important that EMS teams work well together and have a successful partnership when responding to emergency calls.

Day-to-day activities and special events that allow friendships to develop and grow also are going to be important factors for retaining volunteers in EMS. Moreover, because a volunteer's family also is affected by the demands of EMS work, special efforts need to be made to involve spouses and other family members in the EMS organization.

Needs related to friendship and working together can be met in many different ways. EMS managers should decide how they can make volunteers feel they are part of a team and prevent them from becoming isolated and excluded from the group. The manager could either assume or delegate responsibility for seeing that new volunteers have a buddy assigned to help them become part of the group.

At the same time, it must be remembered that the cliques that often form in social groups may make some volunteers feel like outsiders and eventually may lead to their dropping out. The isolated volunteer needs to be integrated into the group.

Symbols of belonging to a group or organization are important reinforcers of allegiance and membership. Uniforms, caps, insignias, or logos on equipment and clothing promote loyalty and a sense of belonging to the group.

Feeling of Satisfaction and Importance

The satisfaction and importance that many EMS volunteers feel as a result of taking charge of an emergency situation and providing lifesaving emergency care is another strong motivational force.

Motivation Tips

- Remember special events in the life of each volunteer, such as the member's birthday or anniversary with the service.

- Provide a physical environment that encourages volunteers to congregate, perhaps for coffee or conversation.

- Personalize your contacts with volunteers and treat them as individuals.

- Determine who works well together, who complements each other, and who could serve as a mentor to new or young volunteers.

Altruism

A powerful reason to join and remain in organizations is the nature of the work itself--providing onscene emergency medical care. EMS work enables the volunteer to feel like he or she is giving something beneficial to others, an altruistic motive. Many Americans are motivated to volunteer as a result of a desire to do something helpful for their community, and the belief that citizens should give something back to society is strong in many communities.

Sense of Achievement and Self-Esteem

EMS volunteers may develop a sense of achievement and self-esteem from being able to do the right thing at the right time. Achievement grows with skill and ability. Individuals who are motivated by achievement need the chance to feel like they are developing expertise and competence as an EMS volunteer.

Successful Performance

Successful performance is a strong motivator for continued volunteer service. The EMS manager needs to communicate to some volunteers the realization that they have been successful by making them aware of their own personal wins and accomplishments. After considering the cultural differences among individuals, the EMS manager can convey a sense of success by acknowledging a volunteer's accomplishments to the organization, in the media, or at special events. By asking volunteers for their advice and ideas about how to get things done, the manager also demonstrates that their accomplishments have been recognized.

Participation in Decisionmaking

It's important for EMS personnel to be able to participate in deciding how work should be organized and who can and should do what. They need an opportunity to:

- Feel that they are doing what they do best.

- Have some ability to decide how things should be done.

- Be challenged by problems that are the right scale for them to solve.

- Get a sense of accomplishment for the work they do.

- Feel that they are improving in their skills and ability.

- Feel that they can manage the tasks assigned.

The demands of a crisis will challenge and energize volunteers if they are prepared for such events and are adequately trained and properly supported. They need the confidence that their support teams and equipment will not let them down.

Following Up After the Application Is Received

Bringing someone on board is not just getting them to fill out an application. Initial followup is the key to getting someone who is interested and has applied for membership to become a volunteer. Feedback information should be provided to new recruits within the first ten days after they apply for membership. The application process should be completed efficiently so that applicants know the results within a month. Potential volunteers can be invited to special events or activities so they have an opportunity to become acquainted with the other volunteers and the way the EMS organization operates.

Post-Application Interview

A special interview should be conducted at the volunteer's home or at the EMS station soon after an application for membership has been received. This interview is the responsibility of the volunteer coordinator or recruitment officer, but it can be delegated to others within the organization.

The recruit interview gives the EMS organization the chance to provide more information about the organization as well as to learn more about the potential volunteer and what he or she expects and is prepared to give. It also may afford an opportunity to explain the social aspects of the organization and the way other family members can enjoy activities or events related to EMS.

The spouse and family of the volunteer should be included in the interview. It is important for married individuals to be able to have their spouses included in the EMS organization's activities whenever possible. The interviewer should explain to the recruit's spouse the level of involvement available to the recruit, the need for specialized volunteers, and the specific opportunities available to the spouse for participating in the activities of the organization.

All or some of the following steps may be included in the post-application interview with each recruit:

1. *Exchange of information* (both written and oral). The interviewer should request the recruited volunteer to share such information as past work experience, interests and skills, reasons for volunteering, and time available.

The interviewer in turn should share with the recruit such information as the nature of the screening process, the organizational purpose, structure, and policies, an overview of volunteer service, the jobs other than EMT that are available, and the general expectations and limitations of volunteers. This is the opportunity for the organization to shine. The interviewer must present a glowing report on the EMS organization; the positive things about the organization should be emphasized. The interviewer who is negative may turn off the volunteer before ever getting started.

2. *Discussion and clarification of the information.* Both the volunteer and the interviewer should be given an opportunity to ask

questions and make sure there is mutual understanding as to the volunteer's job.

3. *Initial screening*. Initial screening as to whether the volunteer should become a member of the organization is mutually accomplished by the volunteer and the interviewer.

4. *Followup interview*. A second, more indepth interview with the officer of the EMS station or activity to which the volunteer is expected to be assigned should be conducted. The officer should discuss the specific type of job for the volunteer, the closeness of the organization, and the relationship of the membership. The officer should be allowed to express both the officer's and organization's expectations for the volunteer.

5. *Matching* (negotiation of responsibilities). EMS managers often forget about matching or placement. Matching of the volunteer is an important aspect of the interview process. Most of the time that a volunteer will spend in the service will not be spent on emergency calls; instead, the majority of time will be spent at the station. The orientation to station life can be a difficult one for the volunteer. There are many things that need to be done around the station, and the volunteer must realize that those responsibilities are as important to the organization as the emergency duties.

The job description for the volunteer's position may provide for a trial period and explain how matching is negotiated when a change in the volunteer's work assignment is needed by the volunteer or the organization. The use of a written job description ensures that the responsibilities of the volunteer and the organization are understood and agreed upon.

The job description should include specific and measurable tasks that can be used to evaluate the performance of the volunteer. Criteria should be established to evaluate the volunteer's performance of the assigned tasks. These criteria for rating volunteers can be as simple as above standard, at standard, below standard, or no activity.

The job description also should include an estimate of the time involved and identification of the supervisor to whom the volunteer reports and is responsible. Unity of command is extremely important to new volunteers because they want to please everyone, but if they are responsible to more than one supervisor, they will become confused and possibly frustrated to the point where they will resign soon after joining the organization.

Top 13 Interviewing Tips

The interviewer can ensure that an interview with a recruited volunteer is effective and productive by using the following tips for successful interviewing:

1. Try to establish mutual trust and understanding. This initial exposure of the potential volunteer to the EMS organization offers an opportunity for the interviewer to engage in good public relations even if a job is not found that fits the volunteer's needs.

2. Let the potential volunteer know that the information shared during the interview is confidential and will not be divulged to others.

3. Practice effective communications: provide information clearly, have written information and visual aids to reinforce oral communications, and listen and check to ensure that he or she is understood and understands the volunteer. Good communication is a two-way exchange.

4. Value each person by remaining nonjudgmental, except in relation to the volunteer's ability to work within the organization in the positions available. Try not to make assumptions.

5. Look for ability and willingness to do a job. Be alert to whether or not the volunteer's values are compatible with the values of the organization.

6. Encourage the volunteer to do the talking by using open-ended questions, such as: "What made you decide on the organization?"

7. Keep the conversation moving without losing sensitivity to the volunteer.

8. Share enough of yourself to put the volunteer at ease. The volunteer wants to know that the people in the organization are real.

9. Be honest about the organization and the job expectations and limitations.

10. Guide the volunteer to avoid overextending. Enthusiasm may cloud the volunteer's ability to make a realistic decision. It's better to start slowly, then add hours and responsibilities after a trial period if the volunteer still wants to make a greater commitment.

11. Include the spouse and family of the volunteer in the interview. It is important for married individuals to be able to have spouses included in the organization's activities whenever possible.

12. Don't talk a volunteer into something he or she really doesn't want to do. Make sure the volunteer understands the commitment necessary in order to remain active with the organization.

13. Look for nonverbal cues. The volunteer's "body language" will help the interviewer understand the feelings of the volunteer.

Avoiding Common Mistakes During the Interview

The interviewer should be aware of the common mistakes made during recruit interviews and consciously ensure that these errors are avoided:

- Talks too much.

- Does not let the volunteer answer the interviewer's questions.

- Does not stay on the topic.

- Does not address the real concerns of the volunteer.

- Does not hear the unasked questions.

- Lacks detailed information.

- Is unable to respond to the volunteer's concerns for safety or time constraints.

Lets prejudices influence his or her judgment.

The Unspoken Concerns of Volunteers

Potential recruits may have a number of concerns about themselves and the EMS organization that they do not express. The interviewer must ensure that these "unasked questions" are addressed:

- What do I really have to do? Can I manage it? Do I have the skill? Can I handle it emotionally?

- How much time will it demand? Is there enough to keep me interested? Will it put pressure on my regular job or family?

- What danger will I be in? What are the risks?

- Who benefits? Why should I do this?

If a waiting list for new volunteers is maintained, the EMS organization should have some system for periodically recontacting the wait-listed volunteers to let them know they are still in line for a position and that the organization is looking forward to the chance to work with them. This followup also provides an opportunity to remove from the list those volunteers who are no longer available for various reasons. To maintain the interest of volunteers on the waiting list, the EMS organization should try to engage them in organizational events and activities, and, if possible, they should be used in some capacity until an EMS post opens up.

Matching Volunteers to Jobs

Successful job performance by volunteers is affected by a number of factors, many of which are beyond the control of EMS managers and supervisors. One thing that managers can do is to ensure that an appropriate match is made between the kind of skills volunteers have and the demands that will be made on them in an emergency situation. Matching is based on the old idea that you can't fit a square peg into a round hole

Mismatches between volunteers' abilities and the jobs they are expected to perform leave everyone unsatisfied. The victim is not satisfied with the quality of service or treatment he or she receives, the volunteer EMT is frustrated by his or her inability to treat the victim properly, and the EMS team is dissatisfied with the impression it makes on the community or hospital personnel. Matching skills and tasks correctly is a key to good performance. This means that managers need to assess the training requirements for volunteers and to think carefully about the assignments they make.

Matching the right person to the job also means that the first volunteer through the door may not be the right person for the EMS job that is being filled. However, a volunteer who is not appropriate for an EMS job may be able to perform other functions for the organization that are useful and important to the overall effort. Many EMS organizations report successful experiences with volunteers who do not make EMS runs but turn out to be critical players on the team. For example, volunteers may have some of the following needed capabilities:

- mechanical skills;
- accounting skills;
- telephone skills;
- purchasing and inventory control skills;
- equipment maintenance and repair;
- public speaking ability; or
- fundraising skills.

Matching doesn't involve just linking the skills and abilities of volunteers with the duties they are expected to perform. Matching also means making an appropriate connection between the personal needs people have for achievement, friendship, and importance, for example, and the kinds of things the EMS organization offers. If the social aspects of participation are important to volunteers and the organization doesn't have many social activities or provide time for social contact, a good match is not likely to occur. The volunteers will not feel satisfied with the return they get for volunteer service and are likely to quit.

EMS managers also can try to match the expectations that individuals have about the demands volunteering will make on them and the actual demands that are placed on them. The assignment of work based on the volunteers' expectations of the workload, rather than distribution of the work evenly among all volunteers, may serve to fulfill the expectations of many volunteers and increase their willingness to serve. Some volunteers will want to give much time and personal energy to the EMS organization, while others will join with the expectation of making a weekly call at most. Tempering demands with expectations may prevent volunteers from feeling that they are overloaded or underutilized.

EMS managers can enhance the match between volunteers and the EMS organization by not promising things they cannot deliver. The number and nature of emergency calls are beyond the control of any manager, although experience may provide some reasonable indication of what can be expected on a weekly or even daily basis.

Finally, matching means that the volunteers should believe that there is a reasonable balance between what they give and what they get back in order to keep them satisfied as members of the EMS organization. Some volunteers may start to think about quitting if they believe they are giving more than they are getting back. Volunteers, like paid staff, want to feel that they are getting a fair exchange for their contribution. This is why it is so important for EMS managers to understand each volunteer's needs and expectations.

Retention

Keeping volunteers in EMS organizations has become more difficult. An EMS organization's efforts to retain its volunteers are as important to the success of an EMS organization as its recruitment program. The best recruitment program is of little value if the organization cannot retain its members. EMS managers must identify and respond to the social, organizational, and job-related conditions that contribute to early or increased turnover if they are to maintain an adequate level of qualified volunteer staff for the organization. There are many reasons why individuals discontinue volunteer service. Today's lifestyles and work patterns mean that many have less extra time for volunteer work in general. Two-career and single-parent families feel sharp time constraints. The changes in volunteers' lives may make it difficult for them to continue to serve for the same length of time in the same capacities as in the past.

The demands within EMS organizations also can have a negative effect on retention. The requirements for becoming and remaining an EMS professional, volunteer and career, are more stringent. Meeting additional training and certification requirements is time-consuming and even costly. Internal conflicts, potential health and safety risks, work-related stress, and lack of confidence in the use and performance of specialized equipment may contribute to early volunteer resignations.

EMS managers need to identify alternative options and develop a strategy for retaining their most valuable resource--the existing staff of trained, competent volunteers. The factors within the control of the manager that contribute to turnover need to be recognized and eliminated if possible. The EMS manager needs to adjust to those factors that cannot be controlled. For example, opportunities for reduced levels of participation could be provided for volunteers whose personal life or job require them to reduce their time commitment to the organization. Volunteers suffering from boredom or stress in their present positions could be reassigned to new or less stressful jobs at least on a temporary basis. Specific retention techniques are described in the *Recruitment and Retention Strategies* unit.

Volunteer Leadership Issues

There are several issues concerning the leadership of the EMS organization that can affect the satisfaction of volunteers and their continuation with the organization: the management style of the organization, the management skills of the EMS manager and the organization's officers, the communication skills of the EMS manager, the image of the organization, and the organizational climate.

Management Style

People want more responsibility for their own work today, both in their places of employment and in volunteer organizations. Rigid and hierarchical leadership styles are increasingly ineffective, but all people expect someone to take charge and act like the leader. Most people want to know what is expected of them. They prefer that policies, rules, and regulations be spelled out, posted, or available for reading.

Volunteers need to feel a sense of control over their own work and responsibility for their accomplishments. They expect respect from others and want to be included in decisionmaking. For example, volunteers could be allowed to set their own goals and objectives for improvement and performance.

Organizing people to work in teams is more common today and many people thrive in that sort of work arrangement. Some with very high friendship needs will need time for conversation and will want to avoid tension and conflict in the work place.

Management Skills

Volunteers may assume that the leaders of the EMS organization at all levels have the requisite skills to serve as managers; however, leaders who are not managers in their private jobs may lack fundamental management skills. Extensive training is required and provided for volunteers who provide emergency medical services; however, training may not be offered to volunteers who provide management services. Volunteers who are dissatisfied with the management abilities of the officers of the organization and their treatment by those officers will be disinclined to remain with the organization. Training for volunteers who are promoted to management positions should be considered as an important retention tool. Management training may be available for free or for a modest fee through local, regional, and state programs provided by EMS academies, colleges and universities, and EMS offices.

Communication Skills

Volunteer members sometimes hear what an EMS manager says differently than the way the manager thought it was said.

Communication patterns differ between women and men and also across cultural groups. Communication includes an unspoken effort to achieve status or friendship. Messages meant to convey instructions can be garbled and misleading to the receiver. Developing good communication skills comes partly from observing reactions to the way the EMS manager speaks: Do they change their posture toward you? Do they grasp the meaning of your instructions?

A variety of techniques to communicate effectively with volunteers may need to be used because they are in and out of the station and seldom all together at the same time. The EMS manager should consider routine meetings, newsletters, written SOPS, bulletin boards, electronic mail, orientation packets, and videotaped presentations.

The Image of the Organization

Each EMS organization develops its own image and spirit. These things are transferred to the community and make an impression on potential volunteers. EMS managers can instill a sense of pride in volunteers as well as a special camaraderie. A positive image affects the way an EMS organization is thought about, and research tell us that people are more interested in belonging to organizations that they think have a positive image and are more prestigious.

Organizational Climate

Each organization has its own personality and a unique climate that develops from the mix of people, tasks, equipment, and external pressures that make up the organization. A critical task for an EMS manager is to define common goals for organizational members and harness the volunteers' energies into achieving those shared goals. Analysis of the organization's climate will help to determine if the EMS manager must work to make the organization a cooperative workplace where people strive to maintain good work relationships.

Where the fundamental expectations of volunteers about what they will get from being a member of an EMS organization are in conflict with their experience, the organization's environment will become increasingly dysfunctional. The EMS manager should think about the importance of matching individuals and organizations for both good performance and retention, To that end, it might be helpful to ask volunteer members the following questions concerning the organizational climate:

- Do you think that our organization offers you the chance to have the kind of volunteer job that you will want in the future?

- To what extent are you made to feel that you are really a part of our organization?

- To what extent have you made social friendships with people you have met through volunteering in our organization?

- How do you feel about the appearance of our organization's stations?

- How do you feel about the appearance of our emergency-response apparatus?

- How do you feel about our organization's image with the public at large?

- When you first volunteered, how well were our organization's policies explained to you?

- Do you think that there is sufficient opportunity for advancement in our organization?

- When you were first contacted, did the people who talked with you about the organization and the opportunities within it describe them fairly and honestly?

- How do you feel about our organization's training program?

- Does the organization keep you informed about its activities and plans?

- How often do you get involved in planning and decisionmaking in our organization?

- How do you feel when you tell people that you are a member of this organization?

- To what extent do you understand just what work you are supposed to do and what your duties are?

- Do you find the work assigned to you challenging and interesting?

- In general, how well do you like your present position?

■ If you were to start again, do you feel you would volunteer for our organization?

■ Do the officers and supervisors on the job set a good example in their own work habits?

■ When you seek information or help on a difficult problem, how likely are you to get the help you need?

■ Are you encouraged to offer ideas and suggestions for new or better ways of doing things?

■ Do you think that your personal problems will be given adequate attention if you brought them to a supervisor's attention?

■ When you are given new duties and responsibilities, how well are they explained?

■ How do you feel about the officer-member interaction?

■ Please tell us any way in which we can improve our organization.

Stress Management

The decision to quit or drop out of an EMS organization is a function of three different kinds of factors that affect a volunteer's satisfaction with his or her membership: concern about the job, conflict with management, and pressure from family. Demands that cannot be managed create a stress level that may become intolerable. Any one factor or a combination of factors may produce enough personal stress to cause someone to withdraw from volunteer service.

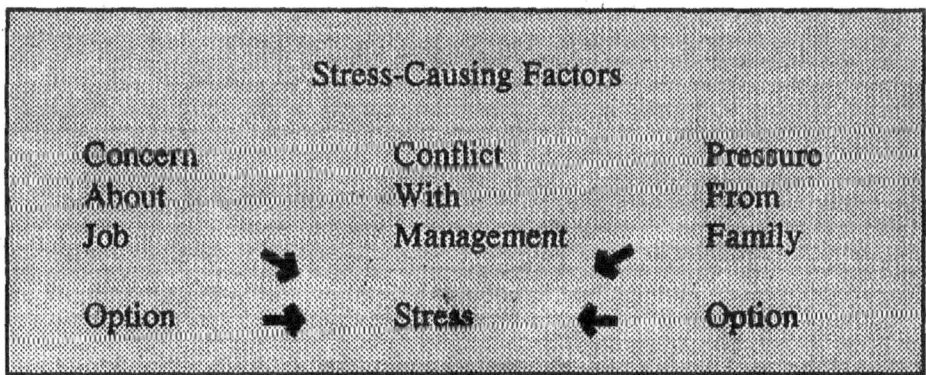

At the same time, the presence or absence of other options for volunteer service may provide the impetus for staying with or leaving the EMS organization. If the other opportunities exist and are as rewarding as participation in the EMS organization, the stressed volunteer may be motivated to join another volunteer organization.

Critical Incident Stress Management

EMS personnel are a unique group of people, but they are also human like everyone else. Those individuals who deliver emergency medical services see other humans when they are most vulnerable as a result of a life-threatening condition or event. Death cannot always be prevented despite the best efforts of EMS personnel. There are so many incidents that an EMS person responds to that could affect the psyche of the volunteer that it is necessary for the EMS organization to have a system in place to care for the emotional well-being of its volunteers. Volunteers who respond to these life-threatening emergencies are subject not only to the stresses of life but also to the effects of being exposed to excessive risks, destruction, and human misery.

The effect of this stress can be seen in the premature resignation of the volunteer or disruption of their health and happiness. If the right steps are taken early enough, the loss of volunteers can be minimized. It is imperative that EMS managers recognize the effects of stress on volunteers and implement a critical incident stress management system.

The U.S. Surgeon General has estimated that 80 percent of the people who die due to nontraumatic causes actually die of stress-related diseases. Stress is not designed by nature to kill, but rather to enhance life. Some stress is helpful and actually essential for a full and productive life. Without stress there would be no change, growth, or productivity.

It is extremely important that a stress management system that provides psychological support services be developed for the EMS organization by someone familiar with the special personalities, operational procedures, stressors, and needs of EMS personnel.

To assure EMS organizations that the very best support services for their personnel are available within the community, it is important that a special psychological support team be created. This team, usually referred to as a critical incident stress debriefing (CISD) team, must be prepared specifically to address the special personalities, job- and family-related stressors, and the support needs of EMS personnel.

If you are not aware of a CISD team, contact your local hospital or mental health organization.

Health and Safety Risks

Concern for our safety is a basic human need. Nevertheless, individuals vary significantly in the amount of risk they are willing to take and the things they consider threatening or dangerous. Good management can't change human nature, but it can minimize unnecessary fears about personal safety. Safety consciousness is good management in its own right.

Individuals who volunteer for EMS may not be overly concerned about personal risks, but they still want the EMS organization to provide valid information about the risks they may face and to take proper precautions for their well-being, particularly in the areas of health, safety, and equipment maintenance.

People in general are increasingly aware of the possibility of contracting a contagious disease. Providing full and understandable information about the nature of health risks EMS volunteers might incur on the one hand and establishing careful procedures to follow to avoid such risks on the other are part of the trust between volunteers and an EMS organization. This trust means providing safety equipment and training for EMS volunteers in the proper use of such equipment to protect them from diseases like hepatitis, AIDS, and tuberculosis. Information about keeping equipment and supplies clean and sterile, when appropriate, and discarding contaminated supplies is part of this training as well. Having safety precautions in place protects EMS volunteers, minimizes concerns for health risks, and enables them to reassure family members that they are not taking unnecessary risks. Good health practices also reduce the spread of infectious diseases among the clientele served.

An EMS manager who is concerned about retention as well as recruitment will enforce the health and safety policies and procedures of the organization. If an EMS volunteer is injured or contracts a disease because rules weren't enforced, the EMS organization may have to be rebuilt in whole or part because of the adverse effect on existing members and the organization's recruitment effort.

Materials concerning the prevention and control of health and safety risks involved in EMS work are available from the U.S. Fire Administration, the Occupational Safety and Health Administration, and the Centers for Disease Control and Prevention.

Equipment Concerns

Equipment must be kept in good condition. Ambulances and other vehicles must be in proper working order, repairs should not be put off, and maintenance schedules should be followed regularly. No one wants to make an emergency run with a vehicle that doesn't operate properly.

The inventory and proper storage of equipment ensures that the right apparatus and gear are available at the moment they are needed. Making a run without the right equipment can frustrate and endanger volunteers and potentially compromise medical treatment.

The regular maintenance of equipment ensures that EMS personnel can rely on the fact that equipment is operable and in the proper location. Inventory control procedures and a checklist for outfitting emergency vehicles are critical for emergency response. The maintenance of safety equipment must be a high priority.

Equipment-related concerns can put stress on an EMS volunteer who is uncertain about how the equipment will perform or whether needed equipment is in the proper location. These concerns may create a sense of not being in control of the situation that many volunteers will find intolerable and may reduce their pride and self-esteem in being an EMS volunteer. For those who are motivated by a desire to help others, failures related to faulty or missing equipment may make them feel useless and frustrated.

The Performance Review as a Retention Tool

Periodic reviews or evaluations of the performance of volunteers by their supervisors are an essential part of EMS management. When handled effectively, these reviews can help in closing the gap between what volunteers do and what the EMS manager needs them to do. They also can increase the likelihood of a volunteer's continued participation in the organization by identifying and responding to any dissatisfaction that volunteers have with job assignments, training, management, and fellow volunteers.

Handling performance reviews effectively, however, is not as easy as it might seem. A number of things can get in the way of an effective review: the design of the evaluation forms, a heavy workload that leaves little time for what seems like extra activities, and the challenge of getting volunteers to accept what the EMS manager has to say. These factors can, but need not, complicate the review process and limit its effectiveness. There are ways to avoid most complications and to make the review process work the way it should.

Defining the Purpose of the Interview

What the purpose of the performance review process is, of course, will depend to some extent on the policies established by the EMS organization, but experience has demonstrated repeatedly that the most effective reviews are those that serve only one purpose: to help the volunteer work closer to his or her potential. Everything that will help a volunteer belongs in the review session unless it can be handled more effectively at some other time.

For example, the correction of a volunteer's tendency to prolong projects certainly would help to ensure that tasks are completed in a more timely manner; however, the correction shouldn't be postponed until review time. It should be made when it is first noticed that the volunteer is not meeting deadlines.

Clarifing the Purpose of the Review to the Volunteer

The volunteers in many EMS organizations do not clearly understand how their work is evaluated or what performance appraisals are all about. If the volunteers are not sure about what's at stake in their reviews and what's really going on, they tend to hold back during evaluation sessions. When pressed, they may become defensive or angry or even more passive than usual.

By clarifying the purpose of the review before it is conducted, the reviewer can avoid a negative response from the volunteers and save a lot of time. If the reviewer and the volunteer being reviewed are absolutely clear about the purpose of the review, it will be much easier to stick to the essential points in the review session and get the important things done.

Conducting the Review as an Exchange of Information

Performance reviews usually involve the completion of an evaluation form that has a "report card" quality to it. The evaluation form is only a tool to help make reviews more objective and more consistent from volunteer to volunteer. Whether it does so depends on how it is designed and used. Even a excellent evaluation form used under ideal conditions is only a means (a basis for discussion) to an end (helping the volunteer).

Report-card-type reviews seldom produce significant or lasting improvements in volunteer performance. Grades may or may not be motivational for children in school, but for most adults, being graded on a job for which you are not paid may provide little incentive to improve.

A more productive approach is to consider the review itself as something that happens between the reviewing officer and the volunteer. It takes the involvement of both sides to make an effective review. There is little chance for improvement unless there is agreement that improvement is needed and possible, and it is unlikely that agreement will be reached without honest and frank discussion. That's why conducting a review as an exchange of information is better than the grade-giving approach. It's not only easier to handle, it's more practical.

Structuring the Interview

The reviewer should organize his or her approach to the discussion of the topics covered in the interview and explain to the volunteer how the interview will be structured. One approach that many managers and supervisors have found useful is the R-A-P review model:

- Review the past.

- Analyze the present.

- Plan the future.

With this approach, the reviewer can trace through time each of the basic topics (such as the job proficiency) from the past through the present and on into the future. This simple chronological sequence is both logical and easy to remember. Consequently, when the R-A-P approach is used, most people have no difficulty in keeping track of where the discussion is going and what progress has been made to date. Another significant feature of this approach is the emphasis it places on the future. Discussion of the past and present is much less important than planning for the future if the goal is really to help volunteers work closer to their potential. Approximately 25 percent of the review time should be spent on the past, 15 percent on the present situation, and a full 60 percent on the plan for the future, which the reviewer and volunteer work out together.

Specifying the Essential Topics to Be Covered

Once the information exchange begins, it may be difficult to stay on target, One way the interviewer can keep the discussion focused is to let the volunteer know in advance the general topics that will be covered in the review. Three essential topics that should be covered are job performance, working relationships with others, and a comparison of what has been accomplished since the last review with what was planned. To establish a sense of two-way communication, which is crucial for an effective interview, the reviewer also should ask the volunteer if there are other topics that he or she wishes to discuss. Volunteers usually want answers to the following questions:

- How am I doing?

- What can I do to improve?

- Do I have a chance for advancement?

- what will be expected of me before the next review?

- How will my work be evaluated during that time?

- What kind of help or attention can I expect from my supervisor?

- What changes are likely in the organization in the months ahead and how will they affect me?

The Exit Interview as a Retention Tool

Many EMS organizations do not conduct exit interviews with volunteers who are leaving the organization because of a perception that the organization does not need that volunteer anymore. In reality, the exit interview can be one of the most important aspects of a retention program if, done in a proper, timely, and professional manner. The information gathered will help the organization in planning for future success.

The purpose of an exit interview is to gather information from each volunteer leaving the EMS organization in order to identify changes or improvements that might be made to reduce turnover of volunteers in the future.

The exit interview may provide the EMS manager with an opportunity to get to the root of a problem or specific reason for leaving and possibly change the volunteer's mind or perception of the situation. It is an excellent time for the leader to determine the specific problem.

There are many reasons for not doing exit interviews, but the primary reason given by most EMS managers is the time required. It can also be an unpleasant task for the EMS manager because the volunteer may try to affix blame for the problem on the manager or another volunteer. This is not the time to place blame but instead a time to determine problems.

Everyone who leaves an EMS organization should be given the opportunity to reply, either in person or in writing to the questions included in the exit interview. A standard questionnaire that can be completed in a face-to-face interview or in writing, should be designed for use in all exit interviews. The following questions should be considered for inclusion in the standard exit interview form.

Sample Exit Interview Format

1. What are your reasons for leaving the organization?

2. What were your most satisfying experiences during your association with the organization?

3. What were your least satisfying experiences during your association with the organization?

4. What was your position in the organization?

5. How would you rate your placement within the organization (poor to excellent)?

6. What training did you receive during your work with the organization?

7. How would you rate the training received (poor to excellent)?

8. What are your recommendations for improving the organization and increasing volunteer satisfaction?

Retention of Career Staff

The turnover of compensated personnel in EMS organizations staffed by combined career-volunteer personnel or career personnel only presents problems and requires solutions that differ from those associated with the turnover of volunteers.

Career personnel leave because of factors that exist both within and outside the EMS organization and the control of its managers. External factors, such as changes in lifestyles, family demands, and other social responsibilities, affect the attitudes of career personnel toward their jobs. Internal factors within the EMS organization, such as job satisfaction, also affect the attitudes of career personnel toward their work and their desire to seek a new position elsewhere. To retain career employees, an EMS manager must keep these internal and external factors in mind and try to make necessary accommodations in the workplace when necessary.

The work performed by a career person, including the routine, daily activities of the job, must be appropriately challenging to the employee. People differ in the extent to which they need change in their jobs and the rate at which they are ready for change. However, a job that is no longer challenging is a major reason why employees become restless and dissatisfied. Employees who feel like they are in over their heads because of the demands of the job also will look for a way out. It is the responsibility of the EMS manager to determine what is appropriate for each individual and to be careful to notice the way people differ from what is thought to be typical in capabilities, interests, and motivations.

Variety is an important part of the workday for many career personnel, but there are certain aspects of their jobs where they want a significant degree of predictability. Volatile relationships between managers and employees or unharmonious relations among employees themselves may make career personnel apprehensive about what to expect on the job and may create an intolerable tension at work. Jobs that become routine and boring may fail to sustain the interest of career employees. EMS managers need to understand the differing needs for predictability and constancy that career personnel have and to adjust the work environment and job assignments to satisfy those needs.

Being an employee is only one of many roles we all perform. The responsibility for small children, aging parents, or ailing relatives is one of many external factors that can place significant demands on a career employee. When the workplace lacks flexibility in both formal and informal rules, employees may feel that they cannot accommodate their other nonwork roles with their role as employee. They may feel pressured to look for a job that is more flexible in terms of work time and time off. The EMS manager should consider the adoption of flexible work schedules; however, to be successful, flexible schedules have to respond to employees' needs and constraints as well as to the work requirements of the EMS organization.

Career employees also may need flexibility as they grow older. They may desire to reduce the physical demands of the job or to make time for other activities that they have postponed. For these career EMS personnel, opportunities for partial employment may be a key factor to their retention. Employees start thinking about how they want to organize their retirement at increasingly early ages.

Partial retirement can be a viable alternative to the loss of a good employee who is ready to do something else or who finds full-time employment too demanding. Partial retirement can be structured in a wide variety of ways, including shorter work days, reduced work weeks or months, or even seasonal employment. Partial retirement options extend the working lives of many employees and are attractive for people in their early fifties as well as those closer to the traditional retirement age. Partial retirement is particularly useful where the work is physically stressful, but also is attractive to employees in a wide range of different occupational groups.

Market factors play an important role in an employee's willingness to remain in an organization. Obviously, in depressed economic periods when other employment opportunities are scarce, employees are likely to remain in an organization, even when they are dissatisfied with the conditions of their job. As soon as the market improves, however, they are likely to leave for other employment. Low turnover therefore cannot be assumed to mean a satisfied workforce. The demand for the skills that EMS personnel have may also increase because of changes in the market. In these cases, the competition for and retention of employees may become a matter of the organization's ability to pay. Most employees, however, will not move from a satisfactory job just for higher pay. Their daily routine, social life, professional identity, and job satisfaction may be more important to them than higher pay. If the pay differential between a career employee's present job and outside opportunities is relatively small, EMS management needs to point out the positive conditions of the job, other than compensation, as a way of retaining employees who have outside opportunities. To the extent that the EMS organization provides opportunities for career development and personal growth, there is a good chance that career personnel can be retained.

Recruitment and Retention Strategies

An effective recruitment and retention program for an EMS organization consists of a package of methods, techniques, or strategies that identify, attract, and keep qualified volunteer members. There is no right or perfect package of strategies, but there are strategies that are more effective than others. The recruitment and retention strategies selected for inclusion in the organization's package or program should be those that have demonstrated effectiveness, can be implemented by the organization at a reasonable cost, and meet the needs of the organization and its membership.

EMS organizations with exemplary recruitment and retention programs were identified by state EMS directors and specialists in EMS recruitment and retention. An analysis of these exemplary programs revealed those recruitment and retention strategies that have been proved effective in actual practice. Either individually or in combination with other strategies. These effective strategies (and the number of the page where the description of each strategy begins) are:

- Annual volunteer recognition event (p. 50)

- Availability of nonoperational opportunities (p. 51)

- Buddy system (p. 53)

- Clearly written job descriptions (p. 54)

- Competitive testing for promotions (p. 55)

- Encouragement of family participation (p. 56)

- Formal recognition system (p. 57)

- Free insurance (p. 59)

- Free meals during long-distance runs (p. 60)

- Free personal equipment and protective clothing (p. 61)

- Free training (p. 63)

- Informal recognition system (p. 65)

- Length of service awards program (LOSAP) (p. 66)

- Mentoring (p. 68)

- Movie theater advertisement (p. 70)

- Multilingual recruitment (p. 72)

- New, well-maintained vehicles (p. 73)

- Open house (p. 75)

- Out-of-town conferences (p. 77)

- Participation-based compensation (p. 79)

- Physical activities (p. 82)

- Piggybacking of recruitment activities (p. 83)

- Print advertisements (p. 84)

- Recruiter incentives (p. 86)

- Stipend (service account) for volunteers who meet minimum weekly participation requirements (p. 88)

- Targeted recruitment (p. 90)

- 24-hour central telephone access by prospective volunteers (P. 92)

- Vacancy announcements (p. 94)

- Volunteer EMT week (p. 95)

- Welcome wagon (p. 97)

- Youth development programs (p. 98)

- Youth education (p. 100)

A description of each of these strategies is provided in the following pages. For each strategy the following information is presented: (1) its applicability (is it appropriate for recruitment, retention, or both?), (2) a general description of the strategy, (3) any special implementation requirements, (4) the strengths of the strategy, (5) any drawbacks or deficiencies of the strategy, (6) its cost, (7) its overall effectiveness as a recruitment and/or retention tool, (8) an example of its use if known (i.e., the experience of at least one EMS organization that has used the strategy successfully), and (9) the name, address, and telephone number of a person who can be contacted to obtain more information about using the strategy.

METHOD	ANNUAL VOLUNTEER RECOGNITION EVENT
APPLICABILITY	Retention
DESCRIPTION	An annual event is conducted to recognize publicly the contributions made by volunteers. The event may be a banquet, dinner dance, or other evening ceremony at which the contributions of volunteers are acknowledged by means of awards and other forms of recognition.
IMPLEMENTATION REQUIREMENTS	A public-recognition event requires careful planning. Arrangements must be made for such requirements as a facility, food service, entertainment, recognition certificates or plaques, event programs and reports, floral displays, souvenirs, gifts, or mementos, and publicity. A coordinator or committee to plan and implement the event will need to be designated.
STRENGTHS	Recognition events provide positive reinforcement of the significant value of the contributions made by volunteers and elicit a sense of personal accomplishment and pride by individual volunteers. Public acknowledgment that the contributions made by volunteers are appreciated will increase the likelihood that they will continue as members of the EMS organization. The events also provide an opportunity to publicize the valuable contribution made by the organization and its volunteers to the community.
DRAWBACKS	The cost of the event may be a problem for some organizations.
COST	There may a significant cost for the recognition events: however, this cost may be covered by inclusion as a planned item in the organization's budget and by contributions of time, products, and services by volunteers and local businesses.
FFFECTIVENESS	Appreciation events are considered to be an essential, valuable retention method.
EXAMPLE	The Odenton Volunteer Fire Company 28 of Odenton, Maryland, is a %-person, largely volunteer organization that provides fire-suppression and basic-life-support/transport services for a community of 25,000. The fire company conducts an annual spring dinner dance to observe the anniversaries of the establishment of the fire company and its ladies auxiliary, acknowledge the achievements of all of its volunteers, and specifically recognize the outstanding achievements of the top ten volunteers in terms of fire responses, EMS responses, total responses, university training hours, practical training hours, and instructional training hours. The printed program for the dinner dance acknowledges the volunteer contributions and provides statistics on the company's activities for the previous calendar year. A two-person banquet committee plans and coordinates the dinner dance.
CONTACT PERSON FOR FURTHER INFORMATION	Charles D. Rogers, Chief Odenton Volunteer Fire Company 28 1425 Annapolis Road Odenton, MD 21113 (410) 674-4444

METHOD	AVAILABILITY OF NONOPERATIONAL OPPORTUNITIES
APPLICABILITY	Recruitment and retention
DESCRIPTION	EMS agencies have a need for many nonmedical services that can be provided by volunteers. These nonoperational services include such diverse activities as equipment maintenance, community relations, public education, training, dispatching, administration, planning, clerical support, photography, mapping, coordination of social functions, and even recruitment. Many potential volunteers do not qualify or do not wish to participate as operational personnel (EMTs or paramedics) because of age, physical disability, or lack of interest, but still are willing or can be motivated to serve as a volunteer in a nonoperational capacity. In addition, operational personnel suffering from burnout or loss of interest often are reinvigorated and achieve satisfaction by performing nonoperational functions. Many EMS organizations provide a range of nonoperational volunteer opportunities as an alternative path for participation by volunteers.
IMPLEMENTATION REQUIREMENTS	The EMS organization must identify the needed services that can be performed by other than operational personnel, define the knowledge and skill requirements for the nonoperational jobs, and expand its recruitment program to target and enlist nonoperational volunteers to perform the needed services.
STRENGTHS	The provision of nonoperational services by volunteers reduces or eliminates the need to use operational personnel to perform those functions instead of or in addition to their life-support functions. The use of nonoperational volunteers enables the EMS organization to draw on a larger pool of potential volunteers to meet its needs. The community image of the organization is enhanced by the variety of citizens able to participate in its activities. Operational personnel whose participation might be decreased or even terminated because of stress or loss of interest can continue with the organization in a nonoperational role.
DRAWBACKS	There do not appear to be any significant drawbacks: however, if a volunteer EMT or paramedic is reassigned to a nonoperational job, the organization may lose its investment in time and money.
COST	There are no significant costs: in fact, the use of nonoperational volunteers may result in a decrease in costs since the financial investment involved in training and qualifying operational personnel is not required.
EFFECTIVENESS	Many staff or support functions of an EMS organization can be performed effectively by nonoperational personnel, and the organization can be successful recruiting volunteers who are not able or interested in serving as an EMT or paramedic. The turnover rate for nonoperational volunteers will be similar to that of operational volunteers. The involvement of nonoperational volunteers is a very efficient use of available volunteer manpower.

EXAMPLE	The Sedona Fire Department of Sedona, Arizona, is a combined career-volunteer organization that provides fire and ALS services from 8 stations for a sparsely populated tourist area covering 126 square miles. The EMS division of the department has a 76 -person staff, including 14 full-time and 3 part-time paid personnel and 59 volunteers. The department makes extensive use of nonoperational volunteers to perform tasks that otherwise would need to be done by operational staff or that it could not otherwise provide. The following departmental needs are filled by appropriately trained volunteers: mapping, photography, maintenance, clerical services, and preplanning. The department's management and operational personnel have given the nonoperational volunteers a high approval rating.
CONTACT PERSON FOR FURTHER INFORMATION	Gary Zimmerman Public Information Officer Sedona Fire Department 2860 Southwest Drive Sedona, AZ 86336 (602) 282-6800

METHOD	BUDDY SYSTEM
APPLICABILITY	Recruitment
DESCRIPTION	The buddy system is a recruitment method that targets the personal contacts of existing members as potential recruits. Each existing member is enlisted as a recruiter and is expected to solicit individuals. with whom the veteran member is familiar, usually friends and relatives, as volunteer members of the EMS organization. The EMS organization may set annual recruitment goals: the usual target is one new volunteer per veteran member each year.
REQUIREMENTS	The EMS organization must formalize the "buddy system" by requiring or at least encouraging veteran-member participation in recruitment setting recruitment goals for the membership, and providing recruitment information (e.g., promotional handout or brochure) to veteran members.
STRENGTHS	The existing members are an excellent source of informed recruiters. They are aware of the requirements for satisfactory job performance, tend to associate with individuals who have similar interests, capabilities, and work schedules, and want the organization to continue as an rganization. They know their friends and relatives as therefore are well situated to assess, contact, and
DRAWBACKS	Good emergency-care providers are not necessarily good recruiters. Some existing members may be reluctant to solicit friends and acquaintances to become members or to take any responsibility for recruitment. The recruitment of buddies may result in "inbreeding" that does not contribute to a diverse membership. If the existing membership does not include females, blacks, non-English speakers, or other minorities, the buddy system may prepetuate organization staffing that is not representative of the area served.
COST	There is minimal additional cost, if any, but the time expended by the entire membership on recruitment will increase.
EFFECTIVENESS	The buddy system will increase the number of recruits, but alone will not meet the volunteer staffing needs of the organization.
EXAMPLE	The Basalt Fire Protection District, headquartered in Basalt, Colorado, is responsible for fire, rescue, and emergency-care services for a low-population, mountainous, 494-square-mile resort area. Basalt Fire and Rescue provides BLS and ALS services through its 28-member, all-volunteer EMT staff: 19 ALS ambulance personnel and 9 first responders. The "buddy method" is the primary recruitment technique used by the department. Each veteran member is expected to recruit at least one new volunteer per year, and most new members are signed up by means of the buddy system.
CONTACT PERSON FOR FURTHER INFORMATION	Steven G. Howard, Chief Basalt Rural Fire Protection District P.O. Box 801 Basalt, CO 81621 (303) 927-3365

METHOD	CLEARLY WRITTEN JOB DESCRIPTIONS
DESCRIPTION	The EMS organization develops and distributes written descriptions for all job classes or positions within the organization for both volunteer and paid personnel. The written descriptions clearly and succinctly lay out for each job the position title, a general summary of the position, the minimum time requirements, the training and certification requirements, the tasks assigned to the position, and the criteria for evaluating performance.
IMPLEMENTATION REQUIREMENTS	The organization management must specify and write job descriptions for all occupational classes or positions within the organization. Personnel publications are available to provide guidance for the development of job descriptions; sample descriptions may be obtained from other local EMS organizations on state-level EMS, fire, or rescue agencies or associations.
STRENGTHS	Well-written, clear job descriptions ensure that prospective and current members of the organization understand the expectations established for each occupational class or position and the basis for evaluating performance: as a consequence, these individuals are better able to decide whether to volunteer as a member or to seek promotion. The written descriptions provide organization managers and supervisors an objective basis by which to establish accountability, evaluate performance, and make initial enlistment and subsequent promotional decisions.
DRAWBACKS	None.
COST	The only cost is the time required for the development or acquisition and periodic review of job descriptions.
EFFECTIVENESS	Written job descriptions contribute to the overall effectiveness of the entire recruitment and retention process, although they alone will not have any significant impact on the process.

METHOD	COMPETITIVE TESTING FOR PROMOTIONS
APPLICABILITY	Retention
DESCRIPTION	Competitive testing is a formal procedure for the ranking and selection of candidates for promotion within an EMS organization on the basis of objective personnel evaluation criteria. Points are assigned to each applicant for an occupational class, slot, or position within the organization on the basis of seniority, written-test results, and oral-interview results. Applicants then are listed in rank order according to total points received. The applicant highest on the ranking list is offered the next open slot for which the applicant has qualified. Formal promotional procedures are administered by a review or merit board established by the governing body for the EMS organization.
IMPLEMENTATION REQUIREMENTS	The promotional procedures prescribed in any applicable state or local legislation must be observed. If state or local law does not apply, the governing body for the organization must authorize and specify the promotional process, including the frequency of the testing (often annually) and the points to be assigned for seniority, written tests, and oral interviews. If permitted by law, the written testing or the entire competitive-testing process can be contracted out to a private testing service or operated internally by the organization through a review or merit board. If the process is not contracted out, the organization must write or purchase objective written tests, administer and grade the tests, assign points, and post the test results. Personal interviews must be scheduled, conducted, and scored: the final ranking lists must be posted.
STRENGTH	The objectivity of the process ensures that the most qualified members are promoted. Favoritism is eliminated because the process is administered in a fair, impartial, and equal manner and is open to all organization members.
DRAWBACKS	Competitive testing removes control of promotions from organization managers, who have had the opportunity over time to know and observe the leadership potential and other qualifications of applicants. The cost of administration of a formal promotional process may be too much for many agencies, especially smaller departments. The testing may be more appropriate for larger agencies that have numerous candidates for open positions.
COST	Contracting out the promotional process can be costly. If the process is administered within the organization, there will be a significant cost in terms of time and money (e.g., purchase of testing instruments)
EFFECTIVENESS	Competitive testing will have a modest positive impact or retention Some members may be encouraged to remain with the organization if fair opportunity to advance to leadership or desirable positions.

METHOD	ENCOURAGEMENT OF FAMILY PARTICIPATION
APPLICABILITY	Recruitment and retention
DESCRIPTION	The family-oriented approach to EMS staffing involves the active recruitment of volunteers from the same household, especially husband-wife pairs. Recruitment efforts emphasize the benefits to both the community and the family of members in the same household working together in a shared volunteer activity.
IMPLEMENTATION REQUIREMENTS	Recruitment activities target families as potential volunteer members.
STRENGTHS	Family-oriented recruitment can attract more volunteers for the same effort. The participation of multiple members from the same family can result in a greater commitment to and a longer duration of participation within the local EMS organization. The involvement of husband-wife teams can lessen the stresses on family life that often result when only one spouse is an EMS volunteer.
DRAWBACKS	Family involvement may result in greater-than-normal membership losses when a single volunteer decides to resign for whatever reason. Families that participate together in EMS will tend to leave together.
COST	No additional cost.
EFFECTIVENESS	Famiily recruitment often can result in family enlistment. Even if only one family member participates, the EMS operation has not expended any additional effort for the recruitment of the single participant.
EXAMPLE	The New Durham Fire Department of New Durham, New Hampshire, provides fire-supression and rescue services for a New England community of 2,300 population. The department is staffed by 30 volunteers, including 8 fire-trained, 2 EMS-trained, and 20 cross-trained personnel. Recruitment is by word of mouth; family and female participation is encouraged. Five husband-wife teams have joined the service.
CONTACT PERSON FOR FURTHER INFORMATION	Terry Jarvis, Captain New Durham Fire Department Box 100 New Durham, NH 03855 (603) 859-3220

METHOD	FORMAL RECOGNITION SYSTEM
APPLICABILITY	Retention
DESCRIPTION	A formal recognition system is the regular practice of an EMS organization to acknowledge the contributions and achievements of Its volunteer members by the presentation of awards to those members who meet eligibility criteria set and publicized internally by the organization. Anything can be used as an award, although they typically include, either alone or in combination, a printed acknowledgment with the recipient's name inscribed (such as a certificate or plaque), an item of value (such as a cash or gift certificate), and a recognition event (such as an awards dinner for members and their spouses). Awards may be presented for any significant achievement: length-of-service milestones, meritorious acts, overall performance (e.g., "EMT of the Year"), level of participation, promotions, and skill-development achievements (e.g., certification, completion of training program). Many organizations believe that as many awards as possible (including humorous awards) should be given as long as the volume of awards does not diminish the value and credibility of the recognition system.
IMPLEMENTATION REQUIREMENTS	An awards committee or task force should be designated to design and administer the recognition system, arrange for the production and presentation of awards, and make recommendations concerning the recognition system to agency management. The type and frequency of the awards to be given, the qualification criteria for receipt of awards, and the process for the nomination of recipients for competitive awards must be specified by the EMS organization through its awards committee and publicized within the organization. The organization must develop and observe an internal procedure for monitoring and acknowledging members who qualify for automatic awards, such as length of service or promotions The process for nominating, evaluating, and selecting the recipients of competitive awards must be implemented on a regular cycle according to the frequency with which the awards are given. Spouses of the recipients of major awards should also be recognized, perhaps with a corsage or companion plaque, in order to acknowledge the important role of family members of volunteer members and to stimulate interest by other spouses. After the recognition system has been announced and expectations created, teh awards must be made as promised in fair and consistent manner.
STRENGTHS	Formal recognition creates interest in reaching acheivable goals, promotes pride in the organization, promotes a competitive spirit among volunteer members, and demonstrates the organization's appreciation for the efforts of it's members.
DRAWBACKS	The value of awards may be diminished if the organization fails to recognize the volunteers who meet the criteria for receipt of an award, gives out undeserved or unearned awards, or requires award recipeints to pay of the costs associated with the award (e.g., tickets to a recognition dinner-dance). If too many awards are given, the recognition system may lose it's meaningfulness and credibility.

COST	There are modest costs associated with the purchase or production of awards (plaques cost $15-20 each): however, some costs may be reduced or avoided. For example, certificates can be prepared with a personal computer and laser printer or by a volunteer member with graphics-design skills.
EFFECTIVENESS	Formal recognition is a effective, low-cost management practice that maintains the interest, morale, and level of participation of volunteer members and thereby contributes to lower and less frequent turnover of members.

METHOD	FREE INSURANCE
APPLICABILITY	Recruitment and retention
DESCRIPTION	The EMS organization pays the premiums for health, disability, workmen's compensation, and/or life insurance to cover its volunteer members.
IMPLEMENTATION REQUIREMENTS	The provision of insurance requires approval by the governing body for the EMS organization and an appropriation by the applicable fiscal body. The types and coverage of and carrier for the insurance to be provided must be selected. Premiums must be paid regularly according to the amounts and schedule prescribed by the carrier. Any state requirements must be observed.
STRENGTHS	Free insurance coverage demonstrates that the EMS organization cares for the health and welfare of its volunteer members. Organization-paid insurance either eliminates the need for members to assume the cost for insurance or allows them to supplement their existing coverage.
DRAWBACKS	Cost.
COST	Insurance plans can be expensive depending on the number of volunteers to be insured and the types of insurance to be provided, although a larger organization may realize lower rates because of the greater pool of individuals to be insured.
EFFECTIVENESS	People respond to financial inducements that remove the need for out-of-pocket expenses that would otherwise need to be paid. Free insurance is a financial incentive that is less attractive than compensation but obviously more appealing than no coverage. Organization-provided insurance coverage contributes positively to the overall recruitment and retention effort.

METHOD	**FREE MEALS DURING LONG-DISTANCE RUNS**
APPLICABILITY	Retention
DESCRIPTION	Free meals are provided by an EMS organization for volunteer members who make long-distance ambulance runs beyond a minimum distance or running time to an emergency-care facility. The meal cost is assumed by the organization by reimbursement of the out-of-pocket meal expense by eligible volunteer members or by maintenance of an organizational charge account at one or more restaurants that may be used by volunteer members.
IMPLEMENTATION REQUIREMENTS	A free-meal Incentive requires an appropriation to pay for the meals and a procedure to provide compensation for the meals.
STRENGTHS	The free meals show that an EMS organization appreciates its volunteers. By taking time to consume a free meal, the volunteers are able to rest, relax, and refresh themselves before the return trip. This time out enables them to wind down and prepare for a safe return.
DRAWBACKS	There is a cost for this retention incentive.
COST	There is a modest cost for this perquisite, which is the product of the number of long-distance runs times the number of staff per run times the average cost per meal. For example, if an EMS organization makes 200 long-distance runs per year, each run Includes a 2-person staff, and the average cost per meal is $5, the total annual cost is $2,000.
EFFECTIVENESS	A free-meal policy alone is not a significant retention incentive, but as part of a package of strategies that demonstrate that an EMS organization values its volunteers members it can contribute to a reduced turnover rate.
EXAMPLE	The Upper San Juan Hospital District, which has its headquarters in Pagosa Springs, Colorado, is a special service district specifically organized to deliver ambulance service for a rugged, rural, 1,800-square-mile area in the southern part of the state. ALS services are provided by a 33-person staff: 3 career members and 30 volunteers who are compensated by call ($15 for local runs, $40 for runs to Durango). The nearest hospital is located in Durango, Colorado, 65 miles from the district headquarters. Volunteer members make about 350 round-trip runs to Durango each year. In addition to their compensation per call, volunteers who travel to the Durango hospital are provided a free meal at one of the two Durango restaurants where the EMT association representing district members has a charge account. The hospital district donates $2,000 annually to the association, and the association uses the unrestricted donation to pay for the meals in Durango. The free meals are a small perquisite that is highly valued by the volunteer members, who appreciate the opportunity to eat for free and rest before returning from Durango.
CONTACT PERSON FOR FURTHER INFORMATION	Eric Schmidt Upper San Juan Hospital District P.O. Box 4189 Pagosa Springs, CO 81157 (303) 731-5811

METHOD	FREE PERSONAL EQUIPMENT AND PROTECTIVE CLOTHING
APPLICABILITY	Retention
DESCRIPTION	Personal equipment and protective clothing are provided by an EMS organization at no cost to its volunteer members, either as standard issue or as an earned reward based on performance, to ensure that they have essential personal gear for undertaking emergency-care or rescue procedures. The free gear may include a fully equipped jump bag (stethoscope, BP cuff, splints, bandages, and other medical supplies), a seperate stethoscope and BP cuff, an oxygen regulator, a glass punch or seat-belt cutter for vehicle extrication work, or protective clothing for rescue work.
IMPLEMENTATION REQUIREMENTS	The personal gear must be purchased and distributed to eligible members. If the EMS organization provides personal gear based on the level of participation of volunteers, it must establish the criteria (runs, in-service hours, points) for receipt of the free gear.
STRENGTHS	The provision of free personal gear ensures that recipeints have the necessary tools to perform their jobs, demonstrates that recipeints are valued members of the organization, and increases the personal confidence and satisfaction of reipeints in performing their jobs. Individually issued personal gear can be maintained in the member's personal motor vehicle and will be available for immediate use.
DRAWBACKS	There is a significant cost to providing free gear. If a free-gear policy has not been in effect previously, the start-up costs for providing free gear to veteran members will require substantial one-time expenditures.
COST	The cost of providinfg free gear is substantial (e.g., $100-$200 for a jumpsuit and $300-$500 for a properly equiped jump bag).
EFFECTIVENESS	A free-gear policy demonstrates that volunteer members are valued members of the organization and that each member's personal capabilities and readiness are important. The policy results in increased pride in and commitment to the organization and, as a result, reduced turnover of volunteer personnel.
EXAMPLE	The Danville Life Saving Crew Of Danville, Virginia, is an all-volunteer, nonprofit organization that provides ALS services for Pittsylvania County, a largely rural area covering 293 square miles and including a population of 83,000, of which 60,000 reside in Danville. The organization Is staffed by 130 volunteers: 25 paramedics, 20 cardiac technicians, and 85 EMT-As. The vehicle fleet includes 6 ALS units, 2 crash trucks, and 2 quick-response vchioloc. The crew provides everything required by a volunteer member at no charge. Crew members receive free training, clothing, and two-way portable radios. The clothing provided includes three coats (lightweight, heavyweight, and dress), shirts, slacks, shoes, and coverall suits. The medical packs stored in the personal vehicles of members may be stocked with crew-provided supplies. The $200,000 annual budget of the crew is funded entirely through public and corporate donations.

CONTACT PERSON FOR FURTHER INFORMATION	Douglas Young Public Information Officer Danville Life Saving Crew 202 Christopher Lane Danville, VA 24541 (804) 792-2739

METHOD	FREE TRAINING
APPLICABILITY	Recruitment and retention
DESCRIPTION	The EMS organization provides free onsite training for its membership and/or reimburses its members for participation in off-site training. The training programs may be basic, entry-level (BLS), advanced-level (ALS), supervisory or leadership, and in-service or continuing-education learning opportunities.
IMPLEMENTATION REQUIREMENTS	Any in-house training requires the planning and coordination of an annual Instructional program: arrangement and scheduling of appropriate facilities: identification, scheduling, and payment of qualified instructional staff: development or procurement of instructor's materials and audiovisuals: identification and procurement of appropriate student materials and expendables: development and Implementation of course evaluation procedures: purchase/ development and maintenance of testing materials and records: and development and maintenance of attendance records. Many organizations maintain an in-house library of audiovisual materials and reference texts. Reimbursement for offsite training must be authorized by the governing body of the organization, reimbursable expenses (e.g., lodging, food, travel, and registration fee) must be specified, and appropriate funding for reimbursement of allowable expenses must be provided.
STRENGTHS	Free training benefits both the participating members and the EMS organization. Free educational opportunities, especially training that is required for membership with the organization or for state certification, are a strong inducement to attract potential volunteers who otherwise would be required to assume the significant cost of mandatory training. Free training at any level increases knowledge and skill levels, improves the capability of the participating members and the organization to provide competent services, increases the confidence of the participating members in performing their duties, reduces the risk of liability of the participating members and organization for any unintended outcomes in providing emergency care, and facilitates education-based advancement within the organization.
DRAWBACKS	Training is expensive, especially offsite training programs, and requires a substantial commitment of time and energy by participating members. Significant training requirements, even if met by in-house training programs, may operate as a disincentive to recruitment and an obstacle to continued retention, especially if the combined training and regular work load are perceived as being too burdensome.
COST	There are significant costs In providing or reimbursing members for participation In training programs. Reimbursable offsite training is more expensive than organization-provided training as the full tuition and other participation costs must be paid. Organization-provided, in-house training Is much less expensive: the major costs are for any outside instructors ($lO-$25 per hour) required, student materials, and consumables. Payment for offsite training may be prohibitive for many departments.

EFFECTIVENESS	Free training is considered to be a very effective recruitment tool by removing the need for assumption of significant out-of-pocket expenses for required training by prospective volunteers. Since many members are reluctant to expend their own monies for advanced or leadership training but are willing to devote their personal time to participate in training, free training eliminates a serious impediment to participation by existing members. Free continuing and advanced training encourages existing members to remain and progress within the organization.
EXAMPLE	Fire Company No. 1 of the Hanover Fire Department in Hanover, Pennsylvania, offers BLS service for a municipality of 14,300 and part of an adjoining township. ALS support is provided by a local hospital. The fire company is a combined career-volunteer operation with two full-time career staff and 30 volunteers. The career personnel staff a single ambulance during weekdays, while the volunteers provide weekday backup and primary coverage at other times. Each volunteer member of the fire department is allocated $750 per year to be expended for any kind of fire- or EMS-related training or education, as well as for protective clothing and necessary equipment. The allowable training expenses include tuition or registration fees and a per diem of $75 per day for meals and lodging: travel costs are not reimbursable. Application for training funds must be made to and approved by the fire commissioner and ambulance captain and be in accordance with guidelines of the state and the firemen's relief association of the fire department. At least 10 volunteer members use their allowance for additional training each year. The source of the funding for the annual allowances is the state-required 2% assessment on the value of fire insurance policies issued in Pennsylvania by out-of-state Insurance companies. The 2% assessment is distributed to and administered by the firemen's relief association of each local fire department according to a formula based on property values and population. The funds may be used for training, purchase of protective clothing and equipment, and death-benefit and retirement plans for volunteers. The Hanover Fire Department receives $40,000 per year in 2% funds for its three fire companies.
CONTACT PERSON FOR FURTHER INFORMATION	James Roth Fire Commissioner Hanover Fire Department 44 Frederick Street Hanover, PA 17331 (717) 637-3877

METHOD	INFORMAL RECOGNITION SYSTEM
APPLICABILITY	Retention
DESCRIPTION	An informal system for the recognition of the important contributions made by volunteer members of an EMS organization is as essential as a system of formal awards to acknowledge member achievements. Informal recognition is accomplishaed in numerous, personal ways by organization managers: saying "thank you" to members for specific activities, involving members in the decision making process of the organization, inquiring about the welfare of the member's family, treating volunteers as equal members of the organization with paid staff, sending notes of appreciation to the member's family, remembering the member's birthday, and celebrating the member's anniversary date with the organization. Informal recognition is accomplished in any way that communicates to the member and the member's family that they are valued as individuals and as key members of the organization family.
IMPLEMENTATION REQUIREMENTS	The only requirement is continous concern, attentiveness, and thoughtfulness by EMS managers in recognizing the role and contributions made by each individual member and the member's family
STRENGTHS	Informal recognition reinforces in a personal way that individual volunteers and their families are important, valued members of the EMS organization and Increases the likelihood that members will wish to continue to serve a friendly, caring organization.
DRAWBACKS	None.
Cost	None.
EFFECTIVENESS	The intangible effects of personal concern and attention to individual members and their families cannot be weighed but should not be underestimated. Most individuals desire to be acknowledged for their efforts and to participate in a caring, friendly, and supportive organization. Informal recognition contributes in an immeasurable way to the continued participation of volunteer members of an EMS organization.

METHOD	**LENGTH OF SERVICE AWARDS PROGRAM (LOSAP)**
APPLICABILITY	Recruitment and retention
DESCRIPTION	A Length of Service Awards Program (LOSAP) is a deferred-compensation or "pension" program for volunteer members of an EMS organization. Monthly benefits are paid to members who qualify after a minimum number of years (typically 5 to 10 years) of active service (i.e., the member becomes "vested"). Benefits are usually based on a fixed monthly dollar amount per year of active service and may be paid when the member retires from the organization at a prescribed retirement age. Benefits to vested members typically may be drawn down at different retirement ages, usually not less than 55, but may be paid at reduced levels at the earlier retirement ages. Benefits are paid automatically to vested members when the members reach any agency-imposed mandatory retirement age. LOSAPs are funded by contributions by the organization or local appropriating body to a self-administered local LOSAP plan or a state or private plan.
IMPLEMENTATION REQUIREMENTS	A feasibility study of the anticipated cost of contribution/compensation alternatives at different retirement ages for a LOSAP must be conducted. A determination must be made as to whether to have a self-administered plan or buy into a state or private plan. The minimum period for vesting, the age(s) of optional or mandatory retirement, the monthly payments (including any payments based on years of service), the basis for earning years of service, and the desirability, amount, and duration of survivor's benefits must be determined. Contributions must be made to the LOSAP plan administrator in the amount and according to the schedule prescribed for the plan. Appropriate monthly payments must be made to vested retired members or their survivors. The timely submission of paperwork is vital to ensure benefits are not jeopardized and are paid in a timely manner.
STRENGTHS	LOSAPs are a strong inducement for volunteers who are older (35 years of age) to enlist with the organization or who have served for some time (8 to 10 years) to remain with the organization in order to qualify for the LOSAP benefits.
DRAWBACKS	LOSAPS require a significant outlay, regardless of the type of plan involved, and may be prohibitively expensive. Self-administered plans require time and expense to administer and involve financial risk in planning investments. LOSAPs do not appear to be very effective in recruitment or retention with younger, shorter-term members.
COST	LOSAPs are very costly ($150 to $200 per month payout premium). Total cost is dependent on the size of the pool of LOSAP participants and the target monthly payout.
EFFECTIVENESS	LOSAPs contribute positively to the recruitment and retention of older individuals, but are less effective with younger individuals.

EXAMPLE	Emergency medical services in Montgomery County, Maryland, a highly populated area northwest of Washington, DC, are provided by the 1,700 career and volunteer members of the 17 independent fire departments and 2 independent rescue squads operating in the county. The Montgomery County Fire and Rescue Commission develops and implements policy, standards, and regulations for fire, rescue, and emergency medical services provided within the county. The Department of Fire and Rescue Services, a division of county government, supports the commission and the fire and rescue corporations and is the employer of all paid fire, rescue, and EMS personnel in the county, Montgomery County provides a LOSAP for volunteer fire, rescue, and EMS personnel that is funded through annuities and general-budget appropriations. Volunteers are entitled to retirement compensation as follows: (1) 25 years of credited service, eligible for retirement at age 60 with $200 per month for life plus $10 per month for each additional year of credited service over 25 years: (2) 15 years of credited service, eligible for retirement at age 65 with $10 per month for each year of credited service: and (3) 10 years of credited service, eligible for retirement at age 70 with $10 per month for each year of credited service. Spouses of deceased members vested under LOSAP are entitled to one-half of the members' retirement entitlement as a survivor's benefit. Years of service are credited on the basis of a point system. The accumulation of 50 points within a year entitles the member to one year of credited service. Points are earned according to a schedule of pointable activities such as participation in drills, training, standbys, sleep-ins, and emergency responses.
CONTACT PERSON FOR FURTHER INFORMATION	Donald D. Flinn Volunteer Coordinator Montgomery County Department of Fire and Rescue Services 101 Monroe Street Rockville, MD 20850 (301) 217-2484

METHOD	**MENTORING**
APPLICABILITY	Retention
DESCRIPTION	Mentoring is the process by which new members of an EMS organization receive the support and guidance of an experienced, veteran member during the first 6 months to 2 years of the new member's participation with the organization. The mentors serve as a volunteer "big brother/sister" or "buddy" to new members during the time, typically a formal probation period, when they face the greatest uncertainties, stresses, and self-doubts relating to their participation as a volunteer.
IMPLEMENTATION REQUIREMENTS	A mentoring program requires the internal recruitment of a cadre of volunteer mentors with the experience, interpersonal skills, and interest to assume the additional responsibility of adviser to one or more new members of the organization. A mentor coordinator to administer the mentoring program needs to be appointed. The mentor coordinator is responsible for the recruitment of new mentors and the matching of mentors with new members according to mentor-member similarities, member needs, and mentor strengths. Guidelines must prepared to define the role of the mentors and the advisory services each mentor is expected to provide. A brief training program for new mentors should eloped for presentation by the mentor coordinator or a veteran mentor.
STRENGTHS	Mentoring exploits an existing resource (i.e., veteran volunteer members) to increase the volunteer retention rate. Veterans who have experienced the same uncertainties, stresses, and self-doubts are well suited to assist new members In avoiding and overcoming the common problems that all new members face. Careful matching of mentors with new members increases the likelihood that their relationship will be successful and that their will be few early dropouts.
DRAWBACKS	There may be an insufficient number of qualified existing personnel who are willing to assume the additional responsibility as a volunteer mentor.
COST	There is little additional cost since mentoring involves the use of existing members who provide extra service as volunteer mentors.
EFFECTIVENESS	Mentoring is a low-cost, effective retention tool that decreases the number of volunteers who resign from the organization during the time when the risk of dropping out is the greatest (i.e., from Initial enlistment through the first year).
EXAMPLE	Emergency medical services In Montomery County, Maryland, a highly populated area northwest of Washington DC, are provided by the 1,700 career and volunteer members of the 17 independent fire department and 2 Independent rescue squads operating in the county. The Montgomery County Fire and Rescue Commission develops and implements policy, standards, and regulations for fire, rescue, and emergency medical services provided within the county.

EXAMPLE (continued)	Each new volunteer or paid member of a fire or rescue company or the Department of Fire and Rescue Services is assigned a veteran member to serve as a mentor during the first two years of the new member's service. In addition, members undergoing ALS training are assigned an ALS-trained preceptor for the duration of the training period. A 2-hour mentor orientation is provided for new mentors; new preceptors receive longer, more formal training. Mentors receive a distinctive "mentor" badge and each month are assigned one to two new members to advise. Mentors are matched with new members according to their similarities, such as sex, personality, and likes/dislikes. A checklist is provided as a guide for mentors to follow during the initial orientation for the new members with whom they have been matched. Each mentor and assigned new member determine the kind of relationship they will have, including the nature and frequency of their contacts. The mentoring program has reduced turnover by increasing the number of recruits who become full members by fulfilling the training requirements and satisfactorily completing the one-year probationary period.
CONTACT PERSON FOR FURTHER INFORMATION	Donald D. Flinn Volunteer Coordinator Montgomery County Department of Fire and Rescue Services 101 Monroe Street Rockville, MD 20850 (301) 217-2484

METHOD	**MOVIE THEATRE ADVERTISEMENT**
APPLICABILITY	Recruitment
DESCRIPTION	A promotional slide or movie clip encouraging volunteer participation in the local EMS organization is run before each showing of a feature movie in local theaters.
IMPLEMENTATION REQUIREMENTS	The managers or advertising representatives for independent or chain movie theaters in the EMS service are are contacted concerning the showing of the promotional ad before each movie on each screen in the affected theaters. A movie clip or slide may need to be produced by the EMS organization for each screen on which the ad is to be displayed. A charge for the showing of the ad will need to be paid if the theater does not display public-service announcements for free.
STRENGTHS	The movie-theater advertising is an inexpensive way to reach a captive audience of potential volunteers and, at the very least, to publicize the service within the community. In addition, the audience reached by the ad is one that apparently has uncommitted evening or week-end time available for recreation and perhaps for volunteer activities.
DRAWBACKS	Some theaters may not accept public-service or commercial advertisements. EMS candidates who express an interest in volunteering as a result of the ad will need to be screened carefully to identify the few who meet the qualifications for acceptance, including time demands and probable longevity of service.
COST	There may be a cost for the production of a slide or film clip if the EMS organization cannot obtain the service for free by a local advertising agency or a volunteer proficient in desktop publishing with a personal computer. Theaters that charge for the showing of the ad may handle the production of a slide upon the submission of rough copy and any photographs to be included.
EFFECTIVENESS	The movie-theater ad approach can contribute as a sufficient low-cost or even no-cost component of an overall volunteer recruitment program. The ad alone, however, could never attract a sufficient number of volunteers to meet the staffing needs of an EMS organization. The ad also serves a public-relations function by reminding the community of the important service provided by the volunteer organization.
EXAMPLE	The Bethesda-Chevy Chase Rescue Squad of Bethesda, Maryland, is a combined volunteer-paid organization with 140 uncompensated volunteers and 8 paid staff that provide ambulance and rescue services to Bethesda, Chevy Chase, and upper northwest Washington, DC. The rescue squad recruits volunteers by means of newspaper and theater ads, billboards, brochures distributed at fairs, stores, and libraries, and contacts during the annual fundraising drive.

EXAMPLE [continued]	The movie-theater advertisements are promotional slides shown in the 10 theaters of a cinema chain located in the rescue squad's service area. The rescue squad arranged for the placement of the ads by contacting the account manager for the cinema chain. For approximately $250 per week, the slide is shown for several seconds, along with other commercial ads, before each feature film at each of the theaters. The ad is displayed about 210 times each week (3 showings of each movie x 10 theaters x 7 days) for a cost of about $1.20 per display. The slide was produced by the cinema chain for free upon submission of rough copy and a photograph by the rescue squad. The rescue squad estimates that 5% to 10% of the volunteers accepted by the squad had applied for membership after viewing the ad: the total number of theater goers who contacted the squad as a result of the ad is not known.
CONTACT PERSON FOR FURTHER INFORMATION	Lewis German, Assistant Chief Bethesda-Chevy Chase Rescue Squad 5020 Battery Lane Bethesda, MD 20814 (301) 652-0077

METHOD	MULTILINGUAL RECRUITMENT
APPLICABILITY	Recruitment
DESCRIPTION	Mulitlingual recruitment is employed in areas where English is not the primary languarge of a significant portion of the population. Recruitment materials and presentations are provided in both English and a non-English language, which usually is Spanish, of a significant segment of the public in order to create a volunteer and career workforce that is representative of the community as a whole.
IMPLEMENTATION REQUIREMENTS	Multilingualism in recruitment requires the replicant of recruitment materials and media in the non-English languages anf the selection and assignment of volunteer and/or paid personnel proficient in the non-English languages to serve as recruiters.
STRENGTHS	Multilingual recruitment promotes a diverse volunteer membership representative of the community. The use of non-English recruitment materials, media, and personnel, coupled with the targeting of the non-English recruitment effort to areas where English is not the primary language, will increase the number of volunteer members of the EMS organization and will contribute to the development of a diverse volunteer membership.
DRAWBACKS	Multilingual recruitment may increase recruitment costs.
COST	Multilingual recruitmentmay increase recruitment costs; however, the costs of replicating recruitment materials and media in a second language may well be offset to an extent by a decreased need to produce as many English-language recruitment materials and media.
EFFECTIVENESS	Multilingual recruitment is targeted recruitment and such as is effective in increasing the involvement of the targeted audience, which in this case is the non-English-speaking population that the EMS organization serves and from which it recruits new members.
EXAMPLE	Volusia County Fire Services, which is headquarters in DeLand, Florida, provides fire, rescue, and nontransport BLS and ALS services for a population of 168,000 residing in the 1,200-square mile unincorporated area of Volusia County. The department is staffed 76 career and 620 volunteer personnel, including 22 career and 6 volunteer paramedics. Twenty-two rescue vehicles, including 4 ALS rescue trucks, operate out of 20 stations. The department serves a multiethnic region with a significant Hispanic population. As a result, firefigthers are required to learn Spanish, and recruitment is conducted in both English and Spanish.
CONTACT PERSON FOR FURTHER INFORMATION	Frank Pocica, Quad Chief. Volusia County Fire Services 123 W. Indiana Avenue DeLand, FL 32720-4619 (904) 736-5941

METHOD	NEW, WELL-MAINTAINED VEHICLES
APPLICABILITY	Recruitment and retention
DESCRIPTION	The EMS organization periodically purchases new ambulances and other vehicles and ensures that they are well maintained and clean.
IMPLEMENTATION REQUIREMENTS	The acquisition of new vehicles requires approval of the governing body of the organization and an appropriation. Design specifications must be prepared and standard equipment-acquisition procedures implemented. A preventive-maintenance schedule for all vehicles must be established and observed. All vehicles must be washed and waxed as needed to ensure they are always clean and presentable.
STRENGTHS	The general public and potential recruits are attached to new, shiny, clean equipment. The vehicles therefore serve as a magnet for inducing prospective volunteers to approach the EMS organization and its personnel and to inquire about membership in an organization that provides attractive, well-maintained vehicles for its membership. The availability and use of such vehicles creates pride, confidence, and enthusiasim among the existing membership.
DRAWBACKS	There is a significant cost in purchasing and replacing organization vehicles. The high cost of vehicle acquisition may limit the ability of some organizations to replace or add vehicles as frequently as desired.
COST	Ambulances and other EMS-related vehicles are expensive; however, since ambulances are required in any event, there are no new initial costs in implementing this strategy. The frequency of replacement of existing vehicles will be the critical factor in determining the ultimate long-term costs for vehicle acquisition and whether the organization can maintain a fleet that is as new as preferred. Any costs for vehicle maintanece and cleansing should already be provided for in the organization's existing budget.
EFFECTIVENESS	Clean, well-maintained, relatively new vehicles have a positive magnetic effect on both present and prospective volunteer members of the EMS organization. The quality and condition of the organization's vehicle fleet contribute in an unquantified way to the overall effectiveness of its recruitment and retention program.
EXAMPLE	The Manchester Volunteer Recue Squad of Chesterfield, Virginia, near Richmond, is an all-volunteer, independent nonprofit organization that provides BLS and ALS services for a population of 115,000. The rescue squad is staffed by 80 senior members (including 34 EMT-As, 16 cardiac technicians, and 3 paramedics), 13 junior members (ages 16 to 21), and 33 auxiliary members. An average of 9 to 10 emergency runs are made by the squad each day. The rescue squad has six ambulances: five ALS vehicles, one BLS vehicle, and one ALS quick-response vehicle. All vehicles are either rechassied or replaced every five years. A new ambulance has been purchased

EXAMPLE (continued)	each year for the past four years. Each vehicle is evaluated periodically by the squad's new vehicle committee, which determines which vehicle should be refurbished or replaced each year. An allocation of $75,000 out of a total squad budget of $210,000 is earmarked for the vehicle improvement program. The squad is supported by public donations received during semiannual fundraising, license tag monies, and auxiliary contributions. By having the highest-quality, best-looking vehicles, the rescue squad believes that the public is impressed and is more willing to contribute to the organization: potential recruits are attracted to the squad: and existing members are motivated to maintain the vehicles.
CONTACT PERSON FOR FURTHER INFORMATION	Kathy Langley Training Officer Manchester Volunteer Rescue Squad P.O. Box 198 Chesterfield, VA 23832 (804) 276-4349

METHOD	OPEN HOUSE
APPLICABILITY	Recruitment
DESCRIPTION	An open house is a public enevt conducted by an EMS organization at its facilities to provide community education that promotes the public health and safety (e.g., injury or fire prevention or use of 9-1-1), to inform the public as to the nature and value of the services provided by the organization, and to attract and recruit potential volunteer members for the organization. The open house typically consists of activities that have public appeal (demonstration of EMS and rescue techniques and equipment and inspection of the organization's facilities and its ambulance, rescue, and fire vehicles) coupled with educational displays, videotapes, and free items for distribution (e.g., posters, brochures, smoke detectors). Attendance at the event by individuals interested in EMS affords an apportunity for recruitment of new volunteers. Recruitment may be accomplished by setting up a designated recruitment table or booth staffed by organizational personnel, making recruitment materials (brochures, sign-up forms) available, and continuous showing of promotional videotape
IMPLEMENTATION REQUIREMENTS	The recruitment component of an open house requires modest preparation. A readily identified recruitment table must be provided; knowlegeable, verbal, and presentable staff must be scheduled to staff the table; and print materials (brochures and sign-up forms) must be made available. If videotapes are used, arrangements must be made for a TV an VCR.
STRENGTHS	The open house enables the EMS organization to realize multiple objectives, such as community education, public relations, and volunteer recruitment, for a single outlay of time and money. Recruitment can be conducted with success on the organization's premises with a receptive population that has demonstrated its interest in the organization's activities by virtue of attendance at the open-house event.
DRAWBACKS	The open house is not a selective recruitment method and may therefore attract individuals who are not well suited to be EMS volunteers.
COST	There is little or no additional cost if recruitment is piggybacked on an open house whose primary, publicized purpose is other than recruitment (e.g., promotion of fire or injury prevention or demonstration of a new ambulance or rescue vehicle)
EFFECTIVENESS	The open-house method is a shotgun approach intended to attract as many people tp the station as possible. A significant number of volunteers can be recruited from the large pool of attendees.
EXAMPLE	The Radnor Volunteer Fire Company of Wayne, Pennsylvania, is a private, nonprofit corporation that provides fire and BLS services for a community of 32,000 located outside Philadelphia. The BLS services are delivered by a 45-person EMT staff (4 paid and 41 compensated per call), of which 25 perform ambulance duty only and 20 are firefighters who also run ambulances.

EXAMPLE (continued)	During the annual fire prevention week in October, the company conducts an open house at its station to promote fire prevention and also to recruit new volunteers. During the two weeks before fire prevention week, the company publicizes the open house by announcements in the local newspaper, by PSAs on local radio, TV, and cable stations, and by brochures distributed during educational visits to all schools in the community. School children are urged to attend and bring their parents. At the open house, attendees can inspect vehicles, observe demonstrations, and receive fire-prevention information. A recruit table staffed by members who will present the best image of the company is set up at the open house to obtain the names of attendees who may be interested in participating as volunteers. Leads obtained during the open house and other fire-prevention week activities are followed up by the company's review committee. Approximately 2,500 attended the most recent open house, and 7 volunteers were recruited during this activity.
CONTACT PERSON FOR FURTHER INFORMATION	George J. Fielden, Assistant Chief Radnor Volunteer Fire Company 121 S. Wayne Avenue Wayne, PA 19087 (610) 687-3245

METHOD	OUT-OF-TOWN CONFERENCES
APPLICABILITY	Retention
DESCRIPTION	The EMS organization pays the cost for the attendance of its members at state and national EMS and fire/rescue conferences. The expenses incident to each conference (registration fee, per diem, travel, lodging) are assumed by the agency for each approved attending member. The agency offers the opportunity to members who meet its eligibility criteria concerning (1) the member's status within the agency (e.g., a minimum of one year of service or regular, nonprobationary membership), (2) the maximum number of paid conferences per member per year (e.g., one paid conference per year), and (3) the priority for approval of attendance if there are more applicants than paid slots available (e.g., preference is given to members whose previous paid attendance at a conference is not the most recent). The organization may require the conference participants to attend all sessions and to provide a written or oral report concerning the conference.
IMPLEMENTATION REQUIREMENTS	The practice of paying for attendance in out-of-town conferences requires the approval of the governing body of the EMS organization and appropriate funding. Criteria for approval of attendance and receipt of reimbursement for eligible expenses must be specified and applied.
STRENGTHS	Agency-paid conference participation improves both the attending members and the EMS organization. Participants obtain new job-related information and learn new relevant skills and techniques; they also have an opportunity to network with their peers and gain insights and ideas from this interaction. The organization also benefits: the attending members will become more motivated and knowledgeable, the information gained by the attendees can be passed along to nonattending members, and the overall morale of the agency will be improved.
DRAWBACKS	The major drawbacks are the cost involved and the loss of the member's time and services for regular duty while attending a conference.
COST	The cost of organization-funded participation in conferences varies according to the location and duration of each conference. The estimated per-member cost for attendance at a 3-day event is $250-$500 (lodging, per-diem, and registration fee) plus travel expenses. The estimated per-member cost for each conference will, of course, affect the number of conference-trips the organization can subsidize each year.
EFFECTIVENESS	The provision of agency-funded participation in out-of-town conferences generally is a positive tool that contributes to continued retention. For most members, attendance at conferences increases their motivation and commitment to the organization. Some members, however, may have no interest in conferences or even resist attending them if the opportunity is offered.

EXAMPLE	The Danville Life Saving Crew of Danville, Virginia, is an all-volunteer, nonprofit organization that provides ALS services for Pittsylvania County, a largely rural area covering 293 square miles and including a population of 83,000, of which 60,000 reside in Danville. The organization is staffed by 130 volunteers: 25 paramedics, 20 cardiac technicians, and 85 EMT-As. The vehicle fleet includes 6 ALS units, 2 crash trucks, and 2 quick-response vehicles. The organization provides everything required by a volunteer member at no charge. Crew members receive free training, clothing, and two-way portable radios. Attendance at conferences is paid for: each member is automatically authorized to attend three state conferences each year. All expenses for participation in specialized, offsite training, symposia, or conferences will be paid for if justified and funds are available. The $200,000 annual budget of the crew is funded entirely through public and corporate donations.
CONTACT PERSON FOR FURTHER INFORMATION	Douglas Young Public Information Officer Danville Life Saving Crew 202 Christopher Lane Danville, VA 24541 (804) 792-2739

METHOD	**PARTICIPATION-BASED COMPENSATION**
APPLICABILITY	Recruitment and retention
DESCRIPTION	A participation-based financial reward program provides a variable economic incentive to attract, encourage the active Involvement of, and retain volunteers for an EMS organization. Compensation is earned according to the level of participation in eligible activities, such as emergency-response runs, in-service training, off-site education, time spent in service or on stand-by, and designated nonemergency functions of the organization. The financial reward program has a schedule of eligible activities by which a fixed amount of compensation or fixed number of incentive points is earned. If compensation per run or call is provided, the financial reward earned may be a uniform amount per call for all members or a variable amount per call based on the employment status of the member (e.g., rank, time in service, pay grade). The compensation scheme also may provide for reimbursement of some or all of the out-of-pocket expenses of participation in off-site education programs. Incentive-point systems typically provide compensation according to a fixed value per point or provide for the distribution of a fixed amount of total compensation to be paid to all members according to each member's percentage of the total incentive points earned by the entire membership Earned compensation is paid on a monthly, quarterly, or annual basis, depending on the type of financial reward program adopted.
IMPLEMENTATION REQUIREMENTS	The method requires the authorization of the compensation scheme by the governing body of the organization and an annual appropriation: development of regulations governing the scheme, including a schedule of compensable activities and, if compensation is to be awarded on a per-activity or per-point basis, the values assigned to each activity; documentation of the compensation or incentive points earned by the volunteers: and a record keeping system for documenting the participation and earned compensation for each volunteer and the payment of the earned compensation.
STRENGTHS	This program provides attractive incentives for the enlistment and participation of volunteers. The variable financial reward program is open to all volunteer members and provides compensation in a fair and equitable manner, i.e., according to each volunteer's contribution to the organization.
DRAWBACKS	The cost of the program may be prohibitive. If volunteers are paid on a hourly basis, the federal Fair Labor Standards Act requires minimum compensation of $4.50 per hour. If an organization has combined career-volunteer staffing, the career members may resent the payments to "volunteers." Animosity may develop between members with little free time (e.g., members with families) and those able to devote considerable time to volunteer work (e.g., unmarried members). It is difficult to discontinue a financial rewards system once started because many members will have joined in reliance on the promise to provide compensation. The provision of compensation gives the appearance that the organization is not a volunteer operation and may make it more

DRAWBACKS (continued)	difficult to obtain donations and support. This approach also may encourage an over response to incidents in order to earn payment or incentive points. It may be more difficult to anticipate and budget for the total annual cost of the financial rewards because the cost will vary according to the number of runs and the members' participation in other compensable activities.
COST	The cost will depend on the number of volunteers, the dollar value of the activities for which compensation can be earned, and the number of compensable activities performed. For example, if an EMS organization has 20 volunteer members who perform an average of 200 compensable activities per volunteer per year with a value of $5 per activity, the annual cost would be $20,000.
EFFECTIVENESS	People respond to financial inducements. Participation-based compensation schemes are effective in increasing the enlistment of volunteers and reducing turnover. They are less expensive than full-time paid staffing but more expensive than uncompensated volunteer systems.
EXAMPLE	The Kettering Fire Department provides fire, rescue, and paramedic (and backup BLS) services for the 61,000 residents of Kettering, Ohio, a suburb of Dayton. The fire department is a combined career-volunteer municipal service staffed by 50 full-time and 120 volunteer firefighters, including 25 full-time paramedics, 10 volunteer paramedics, 25 full-time EMTs, and 37 volunteer EMTs. Two paramedic units are manned around-the-clock by the full-time paramedics, while two ambulances are available for backup BLS service on an on-call basis by the volunteer paramedics and EMTs. The city has a full-time volunteer coordinator to recruit, retain, and recognize volunteer firefighters. The fire department has a volunteer incentive program to provide compensation to volunteer firefighters on the basis of incentive points earned. Each incentive point has a value of $4. Points may be earned for participation in (1) training, (2) emergency responses, (3) company in-service, and (4) management as an officer. Volunteers who participate in in-service training earn one incentive point per hour for training in excess of 16 hours per quarter, with a maximum of 40 hours allowed. Volunteers receive a bonus of $350 for EMT-A certification and $100 for recertification. Volunteers earn two incentive points for timely, full participation in an emergency response, although only 14 total points can be awarded for minor, single-engine responses regardless of the number of responding volunteer firefighters. All active, certified volunteer firefighters share in a company in-service incentive, by which incentive points (maximum of 6 points per day) are awarded quarterly on the basis of the percentage of time a minimum-size volunteer crew for a company Is actually in service. Incentive points also are awarded to officers in recognition of their extra efforts. Incentive rewards average $500 per quarter for each volunteer firefighter for a total annual cost of approximately $240,000.

CONTACT PERSON FOR FURTHER INFORMATION	Joyce Conner Volunteer Coordinator City of Kettering 3600 Shroyer Road Kettering, OH 45429 (513) 296-2433

METHOD	PHYSICAL ACTIVITIES
APPLICABILITY	Retention
DESCRIPTION	Physical activities for all members of an EMS agency are conducted periodically to meet the physical, social, and recreational needs of the members, involve their spouses as important members of the agency's "family," and promote teamwork and camaraderie among members. Typical physical activities include softball, basketball, volleyball, and bowling, Participation in physical activities can be formal (e.g., involvement in a community softball league and/or tournament) or informal (e.g., mixed-double bowling competitions among members and their spouses).
IMPLEMENTATION REQUIREMENTS	Physical activities must be planned and coordinated. A physical activities or fun coordinator or physical activities committee should be appointed to schedule and carry out an annual program of physical activities. Funding may need to be provided for certain activities (e.g., sports equipment, league fees, team t-shirts), although costs can be reduced or avoided through business donations and member contributions.
STRENGTHS	A well-rounded organization takes care of its members' entire needs, including physical exercise and social interaction, and encourages family acceptance of the members' participation as volunteers. Recreational activities increase the members' ties and commitment to the agency, get the spouses of volunteers to "buy in" to the volunteers' active involvement and to participate themselves in agency activities, and fosters a spirit of comradeship among volunteer members. Members are able to select those activities that will be the most fun and the most rewarding.
DRAWBACKS	Physical activities require time to plan and implement. Participation in recreational activities consume additional amounts of free time of the volunteers, although participation is voluntary. Some volunteers may not be interested in or receive satisfaction from the physical activities (e.g., members whose regular jobs are physically taxing).
COST	The cost of recreational functions will vary according to the type of activity. Participation in formal activities may require some expense (uniforms, league fees, sports equipment): however, costs may be defrayed or avoided through contributions from members and through team sponsorship by and donations from local merchants. Informal activities may require some expenses (e.g., volleyball and net), but many activities require no expenditures by the EMS organization (e.g., bowling line charges can be borne by the participants).
EFFECTIVENESS	Physical activities are one important factor contributing to reduced turnover by satisfying the physical, social, and recreational needs of many members, solidifying the commitment of members and their spouses to the organization, and promoting camaraderie.

METHOD	PIGGYBACKING OF RECRUITMENT ACTIVITIES
APPLICABILITY	Recruitment
DESCRIPTION	Piggybacking involves the inclusion of recruitment in other nonrecruitment activities that are provided or paid for by the EMS organization and by other organizations. Examples of piggybacking are (1) the insertion of recruitment messages In regular mailings or newsletters, such as the mail-out of monthly utility bills, (2) the inclusion of recruitment messages in the paid commercial advertising on the packaging of products of cooperating businesses (e.g., pizza boxes), (3) the addition of a recruitment pitch in presentations by EMS organization personnel to community, church, and school groups, and (4) the solicitation of volunteers among participants attending first-aid classes conducted by the EMS organization and open to the public.
IMPLEMENTATION REQUIREMENTS	Opportunities for piggybacking recruitment messages first need to be identified (e.g., preexisting mailings that can be used) and then followed up on.
STRENGTHS	The major value of piggybacking is cost. It is usually free, although a modest charge may need to be paid in some cases.
DRAWBACKS	None.
COST	Free or low cost.
EFFECTIVENESS	Piggybacking is efficient: recruiting costs are avoided or minimized. Whether piggybacking is effective as a recruitment tool depends on the particular event or medium that is piggybacked.
EXAMPLE	The Town of Colonie Emergency Medical Services Department of Latham, New York, is a combined volunteer-paid municipal organization that provides ALS and rescue services to a 60-square-mile area outside Albany with a population of 80,000 residents and an additional 120,000 commuters during weekdays. The department is staffed by 200 volunteers and an additional 50 paid personnel, who provide daytime coverage. The department uses piggybacked recruitment to reach high-payoff volunteers, i.e., individuals who have their roots in the community. As a state-accredited training program, the department provides training that is open to the public and offers an opportunity for volunteer recruitment: in-station first-aid/CPR training for which a fee is charged and BLS training that is funded by the state. First-aid/CPR training is provided off site to community associations and neighborhood groups. The department also advertises for volunteers at no cost in the town newsletter that is mailed periodically to all residents.
CONTACT PERSON FOR FURTHER INFORMATION	Jonathan F. Politis, Director Town of Colonie Emergency Medical Services Department 312 Wolf Road Latham, NY 12110 (5 18) 782-2645

METHOD	**PRINT ADVERTISEMENTS**
APPLICABILITY	Recruitment
DESCRIPTION	Print ads are specially designed, visual advertisements produced for display in any print medium that accepts print advertising, such as newspapers, magazines, newsletters, and billboards.
IMPLEMENTATION REQUIREMENTS	The implementation of every print ad requires concept development, design and layout of reproducible copy, and printing or publication of the ads. The print media in which the ads will be published or displayed must be identified and arrangements made for publication or display, including the cost, timing, and duration of presentation of the ads. Print advertising can be expensive: however, many of the costs can be avoided through in-kind contributions. The ad concept and ad design/layout could be provided at no or reduced charge by commercial advertising firms, freelance advertising consultants, or graphics designers: the concept, as well as camera-ready copy, might be obtained for free from other EMS organizations that have implemented successful print advertising campaigns. Printers and publishers can be requested to waive or reduce their charges for the publication or display of the ads as a public service.
STRENGTHS	Print ads can reach a wide audience of potential volunteers. The targeting of the ads can be focused (as with company or neighborhood-association newsletters) or unfocused (as with newspapers and magazines), depending on the recruitment strategy of the EMS organization. Print ads are also a good public-relations tool to publicize the volunteer-contributed services provided to the community
DRAWBACKS	The cost of producing and publishing/printing print ads can be expensive unless the costs are waived or reduced. The EMS organization may need to allocate additional personnel for answering telephone or mail responses to the print ads, for screening the additional anticipated applicants, and for training and outfitting.
COST	The cost of print ads can be prohibitive for many EMS organizations without a significant budget for recruitment. One EMS organization that carried out a print-ad recruitment campaign reported a cost of about $150 per person who responded to the ad and became a volunteer member.
EFFECTIVENESS	Print advertising enables an EMS organization to compete aggressively for potential volunteers. Print ads, especially those placed in media that enable the organization to target the recruitment message to the types of persons the organization hopes to attract, can be very effective.
EXAMPLE	Eleven independent, volunteer rescue squads of Virginia Beach, Virginia, provide BLS and ALS services without charge for the seaside community of 400,000. The rescue squads operate under the umbrella of the Virginia Beach Department of Emergency Medical Services, a municipal organization that oversees the local EMS system and trains

EXAMPLE (continued)	volunteers for all squads. Virginia Beach reportedly has the largest all-volunteer system in the country with 850 volunteer members. In 1988, when faced with a rising call demand and a shortage of volunteers, the Virginia Beach Rescue Squad spearheaded a 3-week advertising campaign to recruit additional volunteers for all rescue squads. Using advertising concepts developed at no cost by a Norfolk marketing firm, the Virginia Beach squad paid about $15,000 for the production and publication of print advertisements in a local tabloid newspaper, which also ran a cover story on the campaign. The campaign resulted in a flood of calls for information about joining the rescue squads. As a result of the campaign, about 150 volunteers became members of a rescue squad after completing EMT training. The advertising campaign was expanded and repeated in 1990 under the sponsorship of all 11 rescue squads. New advertising concepts were developed by the marketing firm at no cost. The squads paid about $2,000 for the production of newspaper advertisements, $1,500 for the printing of table tents with modified ads for distribution to local businesses, and $1,000 for the printing of the modified ads on billboard posters. A local hospital paid about $10,000 for the publication of the print ads in a local newspaper over a 2-week period, and an outdoor advertising company contributed space on 8 billboards for display of the recruitment posters for 45 days. The newspaper covered the campaign with an inside story. As a result of the campaign, about 175 volunteers became members of a rescue squad after completing EMT training.
CONTACT PERSON FOR FURTHER INFORMATION	Rick Schoew Volunteer Public Relations Consultant Virginia Beach Department of Emergency Medical Services c/o Rick Schoew Marketing 8, Advertising 4020 Silverwood Boulevard Chesapeake, VA 23321 (804) 484-9603

METHOD	**RECRUITER INCENTIVES**
APPLICABILITY	Recruitment
DESCRIPTION	A recruiter incentives program provides inducements for participation by existing volunteer members in the recruitment of new volunteers for the EMS organization. Awards can be given for the most successful recruitments in a year and also for each successful recruitment. The awards must be sufficiently attractive in terms of their desirability that volunteer members will actively engage in soliciting recruits in order to win an award. The award for the most recruitments per year should be substantial, such as a cash bonus of $500 to $1,000 or an expenses-paid trip for two to a vacation spot. The awards for individual recruitments can be more modest, such as $25 to $100 In cash or merchandise.
IMPLEMENTATION REQUIREMENTS	The incentives program must be authorized by the EMS organization's governing body, and an appropriation or budget allocation must be provided to fund the program. The ground rules for the program must be developed and distributed to the membership. The criteria for what constitutes a successful recruitment must be stated unambiguously. These criteria typically include a requirement that the volunteer recruit serve with the organization for a minimum period of time (often one year) and must specify how a successful recruitment resulting from a group recruitment activity (e.g., at a shopping mall or open house) will be credited. Records must be kept as to the number of successful recruitments per member, and arrangements for presenting the recruitment awards, perhaps at an annual dinner, must be made.
STRENGTHS	The strengths of the approach are its ability to involve existing volunteer members as active recruiters, especially members who have effective recruitment skills, its provision of payments to recruiters only for successful recruitments, and its generally reasonable cost per successful recruitment.
DRAWBACKS	The approach may encourage a shotgun approach to recruitment whereby members seek to recruit as many volunteer candidates as possible without regard to their qualifications. Some agencies may not be able to afford to provide sufficiently attractive recruiter incentives. If the only award given is for the highest number of successful recruitments, some members who do not have the time or ability to compete with the most successful recruiters may choose not to participate at all.
COST	The approach involves some cost, which is dependent on the value and number of the awards presented. An incentive program that includes an award for each recruitment as well as an annual top recruiter award will be more expensive than a program that rewards only the recruiter of the year. The cost per recruitment could range from $25 to $300, while the cost for the top recruiter will be more expensive ($500 to $1,000).
EFFECTIVENESS	The recruiter incentives approach is very effective recruitment tool in attracting new volunteers by providing a valuable inducement for existing volunteers to serve as recruiters for the EMS organization.

EXAMPLE	Emergency medical services in Montgomery County, Maryland, a highly populated area northwest of Washington, DC, are provided by the 1,700 career and volunteer members of the 17 independent fire departments and 2 independent rescue squads operating in the county. The Montgomery County Fire and Rescue Commission develops and implements policy, standards, and regulations for fire, rescue, and emergency medical services provided within the county. The Department of Fire and Rescue Services, a division of county government, supports the commission and the fire and rescue corporations and is the employer of all paid fire, rescue, and EMS personnel in the county. A full-time coordinator within the department assists the fire and rescue corporations with volunteer recruitment and retention. The Fire and Rescue Commission established a volunteer recruitment cash award program in 1989 to provide financial incentives for existing volunteer members of the independent fire and rescue corporations and the Department of Fire and Rescue Services to actively participate in the recruitment of new volunteer members. A recruiter award of $100 is paid to each volunteer member who successfully recruits or participates in the recruitment of new volunteers. The recruiter must have actively sought out and encouraged the recruited volunteer to apply for membership in one of the fire or rescue corporations. The volunteer member who receives the most recruiting awards during a calendar year is the "recruiter of the year" and is given a $1,000 cash award with an accompanying plaque. A successful recruitment occurs when the recruited volunteer applies for and is accepted for membership by a fire or rescue corporation or the Department of Fire and Rescue Services, obtains certification as a Firefighter I or EMT-Ambulance, and has served as a volunteer for 12 months. The cash award recruitment program has operated as an effective incentive for volunteer members to enlist new volunteers.
CONTACT PERSON FOR FURTHER INFORMATION	Donald D. Flinn Volunteer Coordinator Montgomery County Department of Fire and Rescue Services 101 Monroe Street Rockville, MD 20850 (301) 217-2484

METHOD	STIPEND (SERVICE ACCOUNT) FOR VOLUNTEERS WHO MEET MINIMUM WEEKLY PARTICIPATION REQUIREMENTS
APPLICABILITY	Both recruitment and retention
DESCRIPTION	A stipend or service account program provides a fixed sum of money that each volunteer member of a local EMS organization is eligible to receive each year by being available for calls for a minimum number of hours per week. For each month the stipend eligibility requirements are met, the volunteer may draw down I/12 of the annual stipend for tax-free EMS-related expenses, such as uniforms, black shoes, emergency lights, and basic emergency equipment. The EMS-related expenses of a qualifying volunteer may be paid either by issuance of a purchase order to the vendor or by reimbursing the volunteer, upon presentation of a receipt, for eligible out-of-pocket expenses. Any balance in the service account at the end of the year is paid to the volunteer in a lump sum as taxable compensation.
IMPLEMENTATION REQUIREMENTS	The method requires an annual appropriation, documentation of the hours of participation by the volunteers, and a financial recordkeeping system for documenting the accumulation of the stipend in each volunteer's service account and payment of eligible expenditures from each account.
STRENGTHS	The stipend is an attractive incentive that gives the volunteer a participation goal to work toward In order to receive the financial reward. The volunteer need not make any out-of-pocket expenditures by using an organization-issued purchase order to obtain EMS-related equipment.
DRAWBACKS	The cost of the stipend may be prohibitive.
COST	The cost is the amount of the annual stipend times the number of qualifying volunteer members plus the cost of administering the stipend program. For example, If the annual stipend Is $500 for an organization with 40 volunteer members, the annual cost for the stipends is not more than $20,000.
EFFECTIVENESS	The method enables the local EMS organization to budget a precise, known sum to obtain the required volunteer services, and everyone receives a fair amount. The cost of a regular "compensation for call" system varies according to the number of calls. This method for compensating volunteers Is less expensive than full-time staffing around the clock, but is more expensive than an uncompensated volunteer system. There appears to be little turnover of volunteers as a result of the use of the stipend method.
EXAMPLE	The Milford Ambulance Service of Milford, New Hampshire, is a 44-person municipal service that operates a combines paid/modified compensated-for-call system in a community of 15,000. Four paid staff provide daytime coverage during the week, while the 40 volunteers provide service on call on nights, weekends, and holidays. Call

EXAMPLE (continued)	members are not required to stay at the station if they live one to two minutes away. The service provides an annual stipend for volunteers who meet a minimum on-call participation requirement per week. The city budgets $20,000 per annum to fund the stipend or service account program for the 40 volunteer members of the ambulance service. Each volunteer is eligible to receive $500 per year by being available for calls for a minimum of 18 hours per week. For each month the stipend eligibility requirements are met, the volunteer may draw down $41 for tax-free EMS-related expenses. The balance of the $500 remaining in the account at the end of the year is paid to the volunteer in a lump sum.
CONTACT PERSON FOR FURTHER INFORMATION	Ron Footit, Chief Milford Ambulance Service 1 Union Square Milford, NH 03055 (603) 673-1087

METHOD	**TARGETED RECRUITMENT**
APPLICABILITY	Recruitment
DESCRIPTION	Targeted recruitment involves the direction of recruiting activities toward those individuals who are most likely to participate in the EMS organization as volunteers and to remain with the organization for a significant duration. These individuals tend to be residents who have their roots in the community: they work in the community, are married and have families, or own their own homes. Nonresidents, such as college students and individuals who work but do not live locally, generally do not have a sufficient stake in the community and are less likely to be long-term volunteers. Targeted recruitment focuses on identifying and enlisting the local stakeholders as volunteers. The recruitment communications channels and messages are designed to reach these stakeholders, including, for example, mailings to homeowners, door-to-door contacts in residential neighborhoods, attendance at neighborhood association meetings, and distribution of brochures at polling places in residential areas.
IMPLEMENTATION REQUIREMENTS	The activities included in a targeted recruitment program will vary according to the opportunities available for reaching local stakeholders.
STRENGTHS	Targeting will result in a greater percentage of high-payoff volunteers being attracted to the EMS organization.
DRAWBACKS	There are no drawbacks to a targeted recruitment program: however, no recruitment program will be 100% successful in attracting high-payoff, long-term qualified volunteers.
COST	The cost will vary according to the specific recruitment techniques employed. Many free or low-cost opportunities are available to recruit the targeted population by piggybacking on existing nonrecruitment activities, such as inclusion of a recruitment flyer with the mailing of monthly utility bills or the solicitation of volunteer participation by individuals who attend first-aid or CPR courses open to the public,
EFFECTIVENESS	Targeted recruitment will increase the number of qualified, low-risk applicants who join the service, reduce the number of applicants who do not qualify for membership or will drop out after participating for less than a year, and reduce the turnover and costs associated with a high turnover rate.
EXAMPLE	The Town of Colonie Emergency Medical Services Department of Latham, New York, is a combined volunteer-paid municipal organization that provides ALS and rescue services to a 60-square-mile area outside Albany with a population of 80,000 residents and an additional 120,000 commuters during weekdays. The department is staffed by 200 volunteers and an additional 50 paid personnel, who provide daytime coverage.

EXAMPLE (continued)	The department looks for high-payoff volunteers-individuals who live in the community and plan to stay, such as residents who have families or are homeowners. The department uses two primary targeted recruitment techniques: (1) as a state-accredited training program, the department provides first-aid/CPR and BLS training that is open to the public, many of whom subsequently become volunteer members: and (2) the department advertises for volunteers In the town newsletter that is mailed periodically to all residents. The department obtains 30% to 40% of its volunteers at the first-aid/CPR courses and another 30% to 40% from the town newsletter mailings.
CONTACT PERSON FOR FURTHER INFORMATION	Jonathan F. Politis, Director Town of Colonie Emergency Medical Services Department 312 Wolf Road Latham, NY 12110 (5 18) 782-2645

METHOD	24-HOUR CENTRAL TELEPHONE ACCESS BY PROSPECTIVE VOLUNTEERS
APPLICABILITY	Recruitment
DESCRIPTION	Central telephone access involves the assignment of a single telephone number to receive inquiries about volunteering as a member of one or more EMS organizations within an area. Telephone access is available 24 hours per day by continuous monitoring of incoming calls and/or by use of an answering machine or voice mail. The volunteer information line is publicized consistently in all media and materials used to communicate with the public or to recruit new members.
IMPLEMENTATION REQUIREMENTS	A single telephone number must be assigned as the access point for receipt of information about volunteer participation. This access point may be the telephone number for an existing line for nonemergency communications or a new, dedicated line to be used solely for incoming calls from prospective volunteers. The use of a dedicated line will more easily permit the use of an answering machine to ensure 24-hour coverage, although agencies with voice-mail capability may add a "volunteer information" option for calls made through a general nonemergency telephone number. To implement central telephone access, the EMS organization must determine whether to use an existing or a new dedicated line, make arrangements for a new line if necessary, and publicize the central-access telephone number in all organization communications, especially recruitment materials. If voice mail is not available, an answering machine with extensive message memory must be obtained and installed. Knowledgeable volunteer staff to respond to incoming calls on a real-time or followup basis must be scheduled.
STRENGTHS	The use of central telephone access, especially when volunteer recruitment is conducted for more than one EMS or fire organization, is an efficient and effective way to ensure that prospective recruits are aware of and promptly receive information about volunteer service.
DRAWBACKS	There are no significant drawbacks: however, once a central number is publicized, the administering organization must staff the line and follow up on all incoming calls.
COST	The only significant cost involved is the price of an answering machine ($100 to $200), although the use of a new, dedicated telephone line will result in additional monthly user charges.
EFFECTIVENESS	A well-publicized central number can facilitate volunteer recruitment.
EXAMPLE	Emergency medical services in Montgomery County, Maryland, a highly populated area northwest of Washington, DC, are provided by the 1,700 career and volunteer members of the 17 independent fire departments and 2 independent rescue squads operating in the county. The Montgomery County Fire and Rescue Commission develops and implements policy, standards, and regulations for fire, rescue, and emergency medical services provided within the county. The

EXAMPLE (continued)	Department of Fire and Rescue Services, a division of county government, supports the commission and the fire and rescue corporations and is the employer of all paid fire, rescue, and EMS personnel in the county. A full-time coordinator within the department assists the fire and rescue corporations with volunteer recruitment and retention. The volunteer coordinator has a dedicated telephone number for 24-hour volunteer recruitment and referral information. This number is advertised in all media used by the Department of Fire and Rescue Services to recruit new members, including cable TV, fixed and mobile bulletin boards, shoppers' guides, flyer inserts, paid newspaper advertisements, brochures, and taxi tops. Calls from prospective volunteers on the dedicated telephone line are responded to by the volunteer coordinator or a member of the recruitment committee. The volunteer information line is serviced by an answering machine to record information requests when calls cannot be handled immediately. All telephone calls are followed up promptly to answer questions and to provide general information about the requirements for participation as a volunteer and information about enlistment with specific fire/rescue companies in the county. A confirming letter is mailed to the prospective volunteer, and a referral form is sent to the specific company affected. Approximately 400 volunteers per year are referred to the fire/rescue companies as a result of inquiries made through the dedicated telephone number.
CONTACT PERSON FOR FURTHER INFORMATION	Donald D. Flinn Volunteer Coordinator Montgomery County Department of Fire and Rescue Services 101 Monroe Street Rockville, MD 20850 (301) 21/-2484

METHOD	VACANCY ANNOUNCEMENTS
APPLICABILITY	Retention
DESCRIPTION	The EMS organization announces to the entire membership the vacancies or openings in each job class, position, or slot as they occur. The vacancies are publicized internally via bulletin-board postings, oral announcements at roll calls and other organization functions, and mailings to the membership, either in a newsletter or separate mailing.
IMPLEMENTATION REQUIREMENTS	The EMS organization must be attentive to the need to inform the membership in a timely manner of all job vacancies as they arise and to use the normal channels for providing important information to the membership.
STRENGTHS	Timely, widespread announcement of vacancies ensures that all members will have an opportunity to apply for and be considered for open slots, eliminates the impression of any favoritism in personnel decisions, and demonstrates that the organization intends to fill open slots in a fair, impartial, and equal manner accessible to all organization members. Open competition for vacancies should result in the advancement of the most competent applicants.
DRAWBACKS	There are no drawbacks to regularly announcing vacancies: however, if the vacant slots announced are not filled in a fair and impartial manner, any positive effect of open announcements will be negated.
COST	The only cost is the time required for the development and dissemination of vacancy announcements, except if separate mailings of the vacancies are sent to the membership.
EFFECTIVENESS	Regular, timely vacancy announcements as a part of the overall retention program may have a modest positive impact on members' decisions to remain with the organization: however, these announcements alone will not have any significant effect. Some members may be encouraged to remain with the organization if there is a fair and open opportunity to advance to leadership or other desirable positions.

METHOD	**VOLUNTEER EMT WEEK**
APPLICABILITY	Recruitment and retention
DESCRIPTION	Volunteer EMT Week Is a concentrated 7-day campaign to recruit and recognize volunteers for one or more EMS organizations, publicize the Important role and activities of an EMS organization, and provide community education that promotes the public health and safety (e.g., Injury or fire prevention or use of 9-1-1). The week's events combine several individual recruitment and retention strategies that could be Implemented independently at any other time (e.g., open houses, TV PSAs, and print advertisements).
IMPLEMENTATION REQUIREMENTS	The specific implementation requirements will vary according to the specific activities and events to be Included in the 7-day campaign. Careful planning and coordination is required in any event, and a Volunteer EMT Week coordinator or coordinating committee must be appointed to plan and manage the week's activities.
STRENGTHS	The week combines the strengths of multiple strategies into a single, focused campaign that ensures there will be greater public visibility and awareness of the need for and rewards to be gained by volunteer membership with the EMS organization. The organization can be selective in choosing the recruitment/retention strategies to be included in the campaign In accordance with its preferences and past experience and success and the available funding and personnel. The costs may shared among several EMS organizations operating within an area by conducting a joint campaign.
DRAWBACKS	The drawbacks of each Individual recruitment/retention strategy will be present in a combined campaign, and the cost may be prohibitive for some agencies with Inadequate funding or personnel to carry out the campaign.
COST	The cost will vary according to the type and scope of the recruitment/retention activities planned for the week and the need for a cash outlay in order to undertake any of the activities.
EFFECTIVENESS	An approach that combines multiple recruitment/retention strategies Into a single campaign, such as Volunteer EMT Week, will be more successful than any single strategy. If the individual strategies used are effective, their use in combination also will be effective, perhaps to a greater degree than normally would be expected because of a synergistic effect (1 + 1 = 3).
EXAMPLE	The Kettering Fire Department provides fire, rescue, and paramedic (and backup BLS) services for the 61,000 residents of Kettering, Ohio, a suburb of Dayton. The fire department is a combined career-volunteer municipal service staffed by 50 full-time and 120 volunteer firefighters, including 25 full-time paramedics, 10 volunteer paramedics, 25 full-time EMTS, and 37 volunteer EMTs. Two paramedic units are manned around-the-clock by the full-time paramedics, while two ambulances are available for backup BLS service on an on-call basis by the

EXAMPLE (continued)	volunteer paramedics and EMTs. The city has a full-time volunteer coordinator to recruit, retain, and recognize volunteer firefighters. An annual Volunteer Firefighter Week is scheduled during the summer to recruit and recognize volunteers and to obtain city and community support for the fire department and its volunteer program. The week's activities are coordinated by the city's volunteer coordinator and conducted in cooperation with neighboring township fire departments. The city allocated $1,000 for expenditures associated with the week as well as the time of the volunteer coordinator. The most recent Volunteer Firefighter Week resulted in the enlistment of 5 volunteer recruits. Volunteer Firefighter Week activities have included (1) a standard slogan or theme used in all recruitment materials ("Part-Time Heroes Needed. Full-Time Rewards"), (2) open houses held at all fire stations, (3) recruitment flyers distributed by a local pizza company with its deliveries, (4) a donated billboard with the recruitment message, (5) specially designed placemats used by a local restaurant chain, (6) a fire and rescue demonstration at which recruitment information is distributed, (7) paid print advertisements, news coverage, feature articles, and supporting editorials in area newspapers, (8) "thank you" messages to volunteers in the bulletins and on the outdoor signs of local churches and on the marquees and in the lobbies of local businesses, (9) gift certificates for volunteers donated by local restaurants, (10) recognition of the volunteer firefighter of the year by a local service club, (11) a special display in the government center, (12) a city proclamation announcing Volunteer Firefighter Week, (13) distribution of flyers publicizing the open houses and the fire and rescue demonstration, (14) radio and TV public service announcements, coverage of events during the week, and interviews with volunteer firefighters, and (15) recruitment inserts included with a local newspaper distributed in targeted areas of the city.
CONTACT PERSON *FOR FURTHER* *INFORMATION*	Joyce Conner Volunteer Coordinator City of Kettering 3600 Shroyer Road Kettering, OH 45429 (513) 296-2433

METHOD	WELCOME WAGON
APPLICABILITY	Recruitment
DESCRIPTION	The welcome-wagon approach recruits potential volunteers by identifying and following up with new residents in a community. Newcomers can be identified by periodically monitoring announcements of real-property transfers in the local newspaper, securing a list of new subscribers for utility services (water, sewage, trash removal) from the appropriate municipal organization, and obtaining the names and addresses of new arrivals from a commercial business that sells such Information for a fee or from a community welcoming service such as welcome wagon.
IMPLEMENTATION REQUIREMENTS	Arrangements must be made to obtain the names and addresses of newcomers regularly from the chosen source (e.g., newspaper, property records office, utilities office, and/or commercial or welcoming service). Mail, telephone, or in-person contacts need to be made by EMS organization staff to explain the nature of the service and need for volunteers and to request participation by the newcomers.
STRENGTHS	The approach is an inexpensive way to reach new arrivals with information about the EMS organization and its need for volunteers.
DRAWBACKS	This approach may not result in great numbers of volunteers and may involve a cost if lists of new residents must be purchased.
COST	The information can be obtained for free if newcomer names and addresses are obtained from public sources: a cost is involved if the information must be purchased.
EFFECTIVENESS	The welcome-wagon method is a shotgun approach and therefore will result In the enlistment of a modest number of volunteers from the large number that are contacted.
EXAMPLE	The West Manchester Township Fire Department of York, Pennsylvania, provides fire, rescue, first-responder, and medical-assistance services for a community of 20,000. The 15-person, all-volunteer EMT staff of the fire department, which has a total membership of 100 active volunteers, is BLS-trained and works closely with the volunteer West York Ambulance Club, which operates from one the fire stations. The department uses a welcome-wagon approach to identify and contact new arrivals to the community about participation as volunteer members. It monitors the property transfer section of the local newspaper to identify new homeowners and Is advised by the municipal utility of new arrivals who have signed up for hook-ups for garbage service. The new residents are then contacted by letter and followup telephone call as to the need for volunteer assistance. This approach has not attracted a significant number of volunteers, but at the same time has not required much effort to implement.
CONTACT PERSON FOR FURTHER INFORMATION	John J. Bierling, Chief West Manchester Township Fire Department 2501 Catherine Street York, PA 17404 (717) 792-3505

METHOD	YOUTH DEVELOPMENT PROGRAMS
APPLICABILITY	Recruitment
DESCRIPTION	A youth program is an organized project sponsored by an EMS organization to provide precareer education and experience for male and female youth 14 to 20 years of age (although many programs are restricted to high-school-age youth). A youth program typically is called a "cadet program" or, if it is affiliated with and approved by the Boy Scouts of America, an "Explorer program." The participating youth receive training (CPR, first responder, BLS), respond with regular members on emergency runs (but are not allowed to enter hazardous environments or to provide emergency medical care), and perform nontechnical and other approved tasks (such as restocking supplies and equipment). Participating youth are assigned a volunteer adviser or mentor to provide guidance and support. Youth programs are intended to provide meaningful extracurricular activities for area youth and to serve as a source of future volunteer members.
IMPLEMENTATION REQUIREMENTS	The youth program and its budget must be approved by the governing body of the organization, a youth coordinator and/or youth committee must be appointed to plan and coordinate the program, a plan must be developed to include a schedule of activities for participating youth, eligibility requirements for participation must be specified and distributed to sources of potential participants (such as high schools), applicants for participation must be screened and accepted, performance guidelines for participating youth and youth advisers must be developed and distributed, and clothing for the participants must be designed, purchased, and distributed.
STRENGTHS	The program provides an opportunity for the participating youth to try out EMS as a possible career at a age when career choices are being considered and enables them to learn first-responder/BLS and leadership skills. The participants contribute to the EMS organization by providing assistance that otherwise would have to be performed by volunteer or paid staff. By conducting what is a preservice EMS training program, the organization develops a pool of trained, experienced EMS personnel, many of whom will be motivated to continue after the program to serve as regular volunteer members, and also performs an important community service by offering meaningful educational and developmental opportunities for area youth.
DRAWBACKS	The program may be too expensive for many departments. At a minimum, the youth participants must be provided minimum clothing and gear. The program must be structured and supervision and direction provided for the participating youth, which means that volunteer or paid staff who otherwise would be engaged in other important organization activities must be diverted to the youth program. Many members may not wish to work as youth advisers or mentors because they do not view an advisory role to be meaningful (e.g., it's "babysitting") or do not want to work with teenagers. Consequently, the

DRAWBACKS [continued]	organization may have difficulty in enlisting a sufficient number of members to serve as youth advisers or may experience a high turnover rate among youth advisers. A youth program will not be a productive source of future recruits in some areas, perhaps rural farming communities, where most youth move away after graduation from high school.
COST	A youth program can be expensive. There are costs for clothing and training youth participants. Members assigned as youth advisers are not available for other organizational activities: as a consequence, the organization may need to incur additional costs to recruit, train, and equip more volunteers.
EFFECTIVENESS	Youth programs are usually a productive source of future recruits, especially where youth are likely to remain in the area after high-school graduation.
EXAMPLE	Volusia County Fire Services, which is headquartered in DeLand, Florida, provides fire, rescue, and nontransport BLS and ALS services for a population of 168,000 residing in the 1,200-square mile unincorporated area of Volusia County. The service is staffed by 76 career and 620 volunteer personnel, including 22 career and 6 volunteer paramedics. Twenty-two rescue vehicles, including 4 ALS rescue trucks, operate out of 20 stations. The service operates a coed Explorer program affiliated with the Boy Scouts of America for high-school youth 14 to 18 years of age. The six stations chartered by the BSA as Explorer Posts provide an opportunity for the current 48 participants to investigate and prepare for a fire-service-related career. Explorers are supervised by volunteer advisers and learn basic firefighting and first-responder skills and often participate in community-related projects, such as fire-prevention and recruitment activities. They are not permitted to enter burning buildings or administer emergency medical care. It costs the sponsoring station approximately $1,800 from its budget to provide uniforms and protective gear for each Explorer. Over one-half of the Explorers subsequently remain with the service as fire-rescue volunteers.
CONTACT PERSON FOR FURTHER INFORMATION	Karen Munson Volusia County Fire Services 123 W. Indiana Avenue DeLand, FL 32720-4619 (904) 736-5940

METHOD	YOUTH EDUCATION
APPLICABILITY	Recruitment
DESCRIPTION	Youth programs are organized activities conducted by the EMS organization to teach youth of various ages how to recognize, avoid, and manage major risks to their health and safety. Educational events can have a single focus (e.g., fire safety) or a comprehensive focus covering all major risks. They can be targeted at certain age groups, such as preschoolers or 5th graders, or at all youth (e.g., children frequenting a mall). They can be provided at a station of the EMS organization, a community facility, or wherever young people are found: schools, malls, summer camps, day care centers, Boys/Girls clubs, YMCAs/YWCAs, or fairs. The education can be a brief intervention, as in the case of a safety-belt promotion at a local mall, or last for one to three days in the case of more comprehensive efforts.
IMPLEMENTATION REQUIREMENTS	The presentation of a youth education program requires careful planning in identifying the risks and target group to be addressed and in arranging for the educational event. An educational program with appropriate audiovisual and print materials must be developed. A facility must be identified and booked. Promotional materials must be produced and disseminated. If the event has limited capacity for participation, a registration process must be organized and implemented. Materials or items to be distributed to participants must be identified, ordered, and purchased, if necessary. Organizational staff and outside speakers who will participate must be identified and scheduled. Arrangements may need to be made for the appearance or use of special vehicles or equipment.
STRENGTHS	Youth education can prevent harmful events that affect youth and their families and ensure that youth are prepared to respond appropriately when a harmful event occurs. Youth programs can serve as a "farm system" for future volunteers: many participating youth will be motivated to join an EMS organization when they are older. These programs are also good public relations. They give public visibility to the organization and its commitment to promote the welfare of the community and its children.
DRAWBACKS	The only drawback may be the diversion of resources of the EMS organization (staff time, vehicles, and some funding) that may be required elsewhere for the direct delivery of EMS services.
COST	The out-of-pocket expense for youth education is modest. Educational materials and media usually can be obtained for free or on loan. The time required for participation of volunteer staff is the major "cost."
EFFECTIVENESS	Youth education can be an effective longer-term recruitment tool, even though recruitment should be thought of as a secondary objective for youth programs. Many participating youth will be motivated to see EMS as a desirable volunteer or even career opportunity. At the least, the participants in youth programs will become community supporters of the EMS organization.

| EXAMPLE | The Mathews Volunteer Rescue Squad of Mathews, Virginia, is an all-volunteer, nonprofit independent organization that provides EMS and rescue services for a very rural tourist area of 8,000 located on Chesapeake Bay. The squad is staffed by 54 volunteers: 4 paramedics, 9 cardiac technicians, 1 shock trauma technician, 33 EMTs, and 7 drivers.

The squad operates "Camp Rescue" twice a year for area youth aged 9 to 13. Each Camp Rescue program runs for one day (9 a.m. to 4 p.m.) for a maximum of 40 participants. The agenda for the program includes both lectures and practical exercises to teach the participants about the major life-threatening risks they or their families may face, such as fire, electrocution, storm and tornado threats, heart attacks, motor vehicle and bicycle crashes, poisons, and drugs, and the actions that can be taken to avoid or respond to these risks. The participants learn about the use of 911, the Heimlich maneuver, opening an airway, rescue breathing, fire safety, bleeding control, poison prevention, burn management, water safety, tornado safety, drug avoidance, recognition and response to heart attacks, electrical safety, street safety, bicycle safety, and the use of safety belts. Each participant is provided refreshments and snacks, a first-aid kit, and a "Camp Rescue" cap. "Camp Rescue" T-shirts may be purchased at cost. Each program costs about $200. Approximately one-third of the participants will join the rescue squad when they are older. |
|---|---|
| CONTACT PERSON FOR FURTHER INFORMATION | Judy Ward
Camp Rescue Coordinator
Mathews Volunteer Rescue Squad
P.O. Box 723
Mathews, VA 23109
(804) 725-2800 |

Annotated Bibliography

Recent publications and articles concerning motivation, recruitment, retention, and volunteerism in EMS organizations are identified and synopsized in this unit. For the most recent information concerning these topics, there are two excellent sources of bibliographies and photocopies of published materials: the Learning Resource Center of the National Emergency Training Center in Emmitsburg, Maryland, and the Florida EMS Clearinghouse in Tallahassee, Florida. Contact information for these sources is included in the *Resources* unit. In addition, the *EMS PIER Manual--Public Information, Education, and Relations in Emergency Medical Services* (September 1994) listed in this bibliography provides further references to publications concerning EMS recruitment and retention,

Motivation

Brown, Rohn M. "Volunteer Programs That Work." *Emergency* (June 1993): 42(2).

> Description of four innovative programs in Virginia to get and keep qualified volunteers: flexible scheduling (40-hour crew), public EMT and CPR training and active media and public information program, college student volunteers, and day care programs.

Buckman, John S. "Motivating Volunteers. " *Fire Engineering* (September 1993):10(2).

> Brief examination of the motivational theories that influence volunteer firefighters as an aid to recruitment, placement, training, supervision, evaluation, and recognition of people. Issues include general observations on human motivation, motivating achievers, motivating affiliators, and motivating power-oriented people.

Colella, Brian. "Volunteer Motivation: Problems and Solutions." *Fire Chief* (October 1992):39(5).

> Description of the major motivational difficulties in the volunteer fire service (maintenance, training participation, qualified personnel for staff and line positions, fire call attendance, and laziness) and possible responses to those problems. Emphasis on an award system, group activities, and assignment of members to teams for nonfireground activities as motivating forces.

Schaubroeck, John, and Ganster, Daniel C. "Beyond the Call of Duty: A Field Study of Extra-role Behavior in Voluntary Organizations." *Human Relations* 44:569(14).

Investigation of the factors influencing voluntary organization members' engaging in a specific extra-role behavior, i.e., participation in a telethon, on behalf of their organizations.

Vineyard, Sue. *Secrets of Motivation--How to Get and Keep Volunteers and Paid Staff* Downers Grove, Illinois: Heritage Arts Publishing, 1991.

Explanation of the connection between motivation and volunteering, the reasons why people volunteer, motivational theories as to what turns people on and off, and ways to keep volunteers and paid staff.

"Watch for the Early Signs of Burnout. " *The Pryor Report 9:7.*

Explanation of the need to recognize signs of burnout in oneself and in employees and to take steps to promote a balanced sense of self throughout the organization. A self-diagnosis for stress is included. Suggestions for supervisors: make well-defined job functions, open lines of communication, seek staff input, and establish an unbiased review system.

Recruitment

Adams, Rich. "Dealing with Population Growth." *Firehouse* 18 (June 1993):14.

Consideration of recruitment/retention and public image as demographics change from rural to suburban.

Barsan, William G. "Emergency Medicine." *JAMA 265* (June 19, 1991): 3115(4).

Report of a study that demonstrates that well-trained paramedics can perform multiple advanced trauma life-support procedures with minimum delay and patient salvage.

Beck, David M. "Help Wanted (We Will Train)." *JEMS* 13 (June 1988):43(7).

Description of the evolution of EMS and its prospects in the future. Good reference source that presents the demographics of volunteer fire service.

Best, Jonathan. "EMS: A System of Response." *Firefighters* News 11 (October/ November 1993):26(2).

  Discussion of whether EMS provided by firefighters or health care professionals offers the best care for a community's money.

Bierwiler, David G. "Recruiting to Reduce Turnover." *Fire Command* (August 1987):42(4).

  Identification of recruiting techniques from application to selection. Emphasis on screening application, physical-agility test, interview, and physical examination.

Burhoe, D.A.P., and Nagele, Richard. "Recruiting in the *80s.*" *Firehouse* (April 1988):68(5).

  Discussion of recruitment issues: reasons for joining volunteer fire departments, identification of problems, and development of recruitment campaigns.

EMS Volunteers for EMS Volunteers. *Volunteer Emergency Medical Systems. A Management Guide.* Blacksburg, Virginia: The Institute for Leadership and Volunteer Development, Virginia Polytechnic Institute and State University, no date.

  Handbook prepared by regional EMS directors in Virginia and Virginia Tech's Institute for Leadership and Volunteer Development to provide management guidance for EMS squads. Management topics addressed include organizations and administration, corporate structure and state licensure, risk management, financial management, fundraising, organizing the membership, effective leadership, planning and evaluation, public relations, and participation in the EMS system and sources of help.

Gilbertson, Michael. "The Volunteer Crisis--Sources and Solutions." *JEMS* (June 1988):6(2).

  Analysis of the reasons for the staffing crisis volunteer ambulance services are facing and recommendation of steps that can be taken to solve the staffing problems.

Haines, Mike. *Volunteers: How to Find Them . . How to Keep Them?* Publisher and date unknown.

  Handbook with ideas, questions, and exercises concerning the planning of a recruitment program, including preparation for recruitment, the procedures to be used once people begin to volunteer, and the methods of recruitment to be used.

Holden, Roger, and Miller, Laurie. "Selecting Officers for Trauma Teams, " *Corrections Today* 55 (August 1993): 190.

> Role of CISDs in reducing the stressful effects of traumatic events.

Hudgins, Edward. "Volunteer Incentives: Solving Recruitment and Attrition Problems." *JEMS* (June 1988):58(4).

> Listing of incentives used to attract and retain volunteer emergency response personnel.

Lynch, Rick. "Targeted Volunteer Recruiting." *Voluntary Action Leadership* (Fall 1990):24(5).

> Eight steps for marketing a product effectively: (1) what is the job that needs to be done? (2) who would want to do it? (3) where will we find them? (4) how will we communicate with them? (5) what are the motivational needs of these people? (6) what will we say to them? (7) who will administer? (8) and how will they know what to do?

Marinucci, Richard A. "Volunteer Recruiting: A Success Story." *Fire Engineering* (August 1988):98(4).

> Discussion of volunteer motivation, marketing, and obstacles.

McCurley, Steve. "Recruiting Volunteers for Difficult Positions." *Voluntary Action Leadership* (Fall 1990):22(3).

> Tips for overcoming the challenges for high skill volunteers: redesigning the job, saturating the market, undertaking a group effort (team volunteering, cluster volunteering), conducting proper orientation, and training up.

McCurley, Steve, and Vineyard, Sue. *101 Tips for Volunteer Recruitment.* Downers Grove, Illinois: Heritage Arts Publishing, 1988.

> Lists of information and tips for conducting an effective volunteer recruitment program.

Morrissey, John. "Here to Stay? Recruiting EMS Volunteers." *JEMS* (February 1993):53(4).

> Tips for establishing a proactive program for recruiting and retaining volunteer members of an EMS organization.

Nestor, Lorretta Gutierrez, and Fillichio, Carl. "Research and Recruitment Strategies: What the American Red Cross Discovered." *Voluntary Action Leadership* (Winter 1992): 15(2).

Report of research by the American Red Cross to discover what motivates volunteers. Thirteen recommendations are made.

New York State Department of Health. *Emergency Medical Services Volunteers--A Recruiting and Retention Guide for Emergency Medical Service Agencies.* Albany, New York: New York State Department of Health, November 1987.

Guidance designed to provide EMS managers and members who serve as recruiters with a step-by-step approach to finding, enrolling, and keeping volunteers. Six steps to successful recruitment are recommended: (1) analyze and plan, (2) advertise and promote, (3) prospect for potential recruits, (4) contact and interview prospects, (5) enroll and follow up, and (6) retain.

Nordberg, Marie. "The Case of the Missing Volunteer." *Emergency Medical Services* 18 (June 1989):13(g).

Outline of societal factors affecting the rate of volunteerism and description of recruitment and retention strategies to deal with shortages.

Perez, Carlos, *et al. EMS PIER Manual--Public Information, Education, and Relations in Emergency Medical Services.* Report No. FA- 151 /September 1994. Washington, DC: National Highway Traffic Safety Administration, U.S. Department of Transportation, and U .S. Fire Administration, Federal Emergency Management Agency, 1994.

Guidance directed to all types of EMS services for creating, maintaining, and enhancing a public information, public education, and public relations program, Topics addressed include starting a PIER program, selling a PIER program, networking, funding a PIER program, media relations, legal issues, onscene PIER, public relations, measuring success, and resources for obtaining help

Perkins, Kenneth B., and Weiderhold, Robert. *Volunteer Firefighters in the U.S. : A Summary of Social Characteristics and Commitment.* National Volunteer Fire Council, no date.

Statistics about volunteer firefighters.

Perkins, Kenneth B., and Wright, Terry. *Characteristics and Survival Potential of Volunteer Nonprofit EMS Corporations: A Virginia Sample.* Blacksburg, Virginia: The Center for Volunteer Development, Virginia Polytechnic Institute and State University, February 1989.

   Description of the findings of research concerning the characteristics of volunteer nonprofit EMS personnel in Virginia and their perceptions of problems faced by volunteer EMS squads in the state and a discussion of the policy implications suggested by the research that could affect the survival of volunteer EMS corporations.

Skeen, D. Trace. "Public/Private Sector Partnership: Myth or Reality." *Firefighters* News 11 (October/November 1993):58(4).

   Description of ways to identify and integrate public/private services and needs.

Snook, Jack W., and Olsen, Dan C. *The Volunteer Firefighter: A Breed Apart.* Fort Collins, Colorado: Emergency Resource Inc., 1989.

   An instructional package, including videotapes and printed materials, designed to provide an overview of the history of volunteerism, the information and skills necessary to determine the need for volunteers and plan for new members, an understanding of the recruitment process, the elements of a successful volunteer training program, leadership skills and techniques to apply in a volunteer environment, ways to motivate volunteers, and ideas on incentive programs.

Swan, Thomas H. "Recruiting EMS Volunteers." *JEMS* 13 (June 1988):51(4).

   Discussion of challenges administrators face in meeting the need for EMS volunteers. Focus is on volunteer utilization, recruitment, and training.

Sylvia, Dick. "Tapping Your Wealth of Volunteer Resources." *American Fire Journal* (August 1987):22(2).

   Advantages that a volunteer fire department has in bringing a variety of outside skills to the station house. Discussion of the need for good trainers and educators and for a diversity in disciplines represented.

Thorn, Don. "Could a PR Firm Help *You?*" *Fire Chief* (September 1987):46(2).

   Worthington, Kentucky, fire department's use of a PR firm. Recommendations made for defining a department's need, obtaining recommendations, and selecting a PR organization that will consider departmental needs seriously and develop an appropriate strategy.

Thompson, Justin J. *Recruitment and Retention of Volunteer Firefighters.* May 18, 1992.

Report of data concerning volunteers and their employers (good background).

Virginia Department of Health, Office of Emergency Medical Services. *Looking Your Best, Getting the Best, Keeping the Best. Virginia EMS Recruitment and Retention Action Kit.* Richmond, Virginia: Virginia Department of Health, no date.

Package of materials to aid EMS organizations in Virginia in planning and implementing an effective EMS recruitment and retention program.

Virginia Department of Health, Office of Emergency Medical Services. *Tune Into the EMS Support Team Network. Supplemental Materials to Help Start Your EMS Support Team.* Richmond, Virginia: Virginia Department of Health, no date.

Guidebook for recruiting and using volunteers to serve as nonoperational support staff for an EMS organization.

Wanous, John P. "Effects of a Realistic Job Preview on Job Acceptance, Job Attitudes, and Job Survival," *Journal of Applied Psychology 58:327(6).*

Report of a field experiment conducted in a telephone company to assess the effects of a realistic job preview versus an unrealistic (i.e., "traditional") preview.

"What if You Called and Nobody Came?" *Size Up* (October 1988):8(2).

Flyer for recruiting volunteer firefighters: benefits, training, apprentice firefighters, need, and time.

Whiting, Mark D. "Lasting Impressions." *Emergency* (June 1993):44(6)

Pragmatic suggestions for rural EMS agencies in establishing and maintaining proactive media relations. Coverage includes the reasons for using the media, ways to get started in developing relations with the media, and ways to maintain a continuing relationship with the media.

Williams, Robert A. "The Future of the Fire Service." *Firehouse* (April 1989):72(4).

Benefits of establishing contacts with youth groups.

Wilson, Marlene. *How to Recruit Today's Volunteers.* Boulder, Colorado: Volunteer Management Associates, 1983.

Training package consisting of a videotape and related materials that provides information on demographics, motivations for volunteering, recruitment tips, and keys to volunteer retention.

Retention

Cordes, Cynthia L., and Dougherty, Thomas W. "A Review and an Integration of Research on Job Burnout." *Academy of Management Review* 18 : 621(3 6).

Review of the literature on burnout and presentation of a conceptual framework to improve the understanding of burnout.

"Critical Incident Stress Debriefing." *Commish* (April 1992):3(3).

Discussion of how firefighters handled stress prior to the 1980s. Identification of two ways to assist departments in dealing with stress: CISD and employee assistance programs.

El Sanadi, Nabil. "High Performance EMS." *Firehouse* (May 1993):68.

Discussion of the elements of high-performance EMS systems: personnel, structure, and equipment and supplies.

Fullerton, Carol S.; McCarroll, James E.; Ursano, Robert, J.; and Wright, Kathleen M. "Psychological Responses of Rescue Workers: Fire Fighters and Trauma. " *American Journal of Orthopsychiatry 62* (July 1992): 371(8).

Examination of the psychological responses of two groups of firefighters following the performance of rescue work.

Goldfarb, Bruce. "Under Pressure." *American Medical News* (November 16, 1992):23(2).

Discussion of burnout and its impact on New York City EMS workers and methods for dealing with stressed personnel: psychological screening to weed out the most fragile applicants and the use of CISD and EAP programs to help workers deal with stress.

Goldfeder, William. "Retaining and Recruiting Members." *Fire Engineering* (May 1992): 10(3).

Presentation of effective activities that volunteer departments may use in retaining active volunteers and recruiting new members.

Herman, Roger E. "Challenge of the Future: Keeping Good People. " Ambulance Industry Journal (September/October 1992):8(3).

Suggestions for retaining employees in the prehospital care industry: find and hire good people, keep good people, create a positive working environment, build positive relationships, help people grow, and learn why good people work for the organization.

Kipperman, Stephanie. "Training Volunteers for Success." *Leadership* (July/ September 1993):20(2).

Goals and objectives, content and focus, and means and techniques of volunteer training programs.

Leibowitz, Zandy B.; Schlossberg, Nancy K.; and Shone, Jane E. "Stopping the Revolving Door. " *Training and Development Journal* 45:43(4).

Innovative strategies for new employee orientation. Three models--self-managed peer groups, rolling out the red carpet, and learning the big picture--that treat new employee adaptation as an ongoing empowerment process are profiled.

Lipp, John L. "The Top 25 Methods to Retain Volunteers." *Voluntary Action Leadership* (Spring 1992):25(2).

List of 25 methods to retain volunteers. Best method is to engage entire paid staff in the process.

Lynch, Richard. "Designing Volunteer Jobs for Results." *Voluntary Action Leadership* (Summer 1983):20(4).

Discussion of the four most important principles for designing a rewarding job for volunteers.

Macduff, Nancy. "Retention: The + and - of Volunteers." *Voluntary Action Leadership* (Winter 1992): 19.

Role of paid staff in affecting volunteer retention rates. Advantages and disadvantages of volunteers. Five questions that should be asked: (1) who volunteers? (2) who doesn't volunteer? (3) why don't certain people volunteer? (4) why do current volunteers give their time? and (5) what benefits arise from volunteering?

MacKenzie, Marilyn. *Dealing With Difficult Volunteers.* Boulder, Colorado: Volunteer Management Associates, 1988.

Explanation of the volunteer retention cycle, the types of difficult volunteers, and principles for dealing with difficult volunteers.

Martin, Kathleen R. "Pulling Together to Cope With the Stress." *Nursing* (May 1992):38(4).

Discussion of the Critical Incident Stress Management (CISM) program and the seven phases of debriefing (introductory, fact, thought, reaction, symptom, teaching, and closure).

Matin, Scott A., and Lester, David. 'Attitudes of Emergency Medical Service Providers Towards AIDS." *Psychological Reports 67 (* 1990) : 1314.

Report on a study of 19 mobile, intensive care paramedics and 21 RNs scored concerning their attitudes towards AIDS, their treatment of AIDS patients, and the effect of AIDS on their continuing in emergency-care operations.

McLeod, John. "Stress: How to Recognize It and What to Do About It." *Fire and Rescue* (January 1992):22.

Description of the reliable, established strategies that fire brigades can use to minimize the negative effects of stress problems: self-help, organizational change, counseling, and Critical Incident Stress Debriefing (CISD).

Miller, Lynn E.; Powell, Gary N.; and Seltzer, Joseph. "Determinants of Turnover Among Volunteers." *Human Relations 43 : 901(17).*

Examination of the causal sequencing of attitudes, personal situations, and behavioral intentions as determinants of turnover among hospital volunteers.

Nordberg, Marie. "Is Stress Killing New York's EMTs?" *Emergency Medical Services* 22 (July 1993):50(4).

Report of a study of New York EMTs' stress following six EMT deaths within an 8-month period.

Paulsgrove, Robin F. "Recruiting and Retaining Fire Department FPEs." *NFPA Journal* 87 (January/February 1993):58(6).

Recommendations made in response to turnover in a Texas fire department: revising advertising procedures, introducing career ladders, developing comprehensive orientation programs, and expanding roles.

Pendleton, Michael; Stotland, Ezra; Spiers, Philip; and Kirsch, Edward. "Stress and Strain Among Police, Firefighters, and Government Workers: A Comparative Analysis." *Criminal Justice and Behavior* 16 (June 1989):196(15).

Comparison of matched samples of police officers, firefighters, and other municipal workers with regard to self-reports of stress and various mental, physical, and behavior problems reflecting strain resulting from stress.

"Postponing the Advent of Paid Firemen." *New Haven Register*, January 17, 1989:8.

The contribution of pension plans to the retention of volunteer firefighters.

National Volunteer Fire Council. Communications Committee. "White Paper: Retention and Recruitment in the Volunteer Fire Service." March 1993.

Discussion paper for a Retention and Recruitment Workshop held in Emmitsburg, Maryland, on March 27-28, 1993.

National Volunteer Fire Council. *Retention and Recruitment in the Volunteer Fire Service--Problems and* Solutions. Report No. FA-138/September 1993. Emmitsburg, Maryland: Federal Emergency Management Agency, U.S. Fire Administration, August 1993.

Report on the general nationwide problems and solutions concerning recruitment and retention in volunteer fire and rescue services.

Selig, Suzanne M., and Borton, Danny. "Keeping Volunteers in EMS." *Voluntary Action Leadership* (Fall 1989): 18(3).

Analysis of information about EMS personnel (volunteers and paid) in a Michigan county and their satisfaction/dissatisfaction with EMS work.

"Stress Management in the Aftermath of Disaster." *Straight Tip 3:6(2)*.

Presentation of excerpts from the booklet *Prevention and Control of Stress Among Emergency Workers* for planning for postdisaster stress management.

Swan, Thomas H. "Keeping Volunteers in Service." *JEMS* (June 1988):55(3).

Recommended techniques for retaining experienced volunteers in EMS agencies by meeting important psychological needs.

Swan, Thomas H. "Volunteers on Duty." *Emergency* (September 1994):68(2).

Importance of formally scheduling EMS volunteers for specific duty shifts.

Taylor, Vickie Harris. "Firefighters: Stress and the Psychological Profile." *Fire and Rescue* (January 1992):20.

Discussion of recurring general psychological characteristics of firefighters: obsessive-compulsive, histrionic, action-oriented, control, family-oriented, instant gratification, highly dedicated, or risk taker.

U. S. Fire Administration. *Emergency Medical Services Management Resource Directory*. Report No. FA-119/November 1992. Emmitsburg, Maryland: USFA, 1992.

Quick-reference document to assist the EMS practitioner or scholar in identifying or researching EMS resources, including pertinent organizations and associations, computer databases, and current reference materials, Directory includes a listing of contacts intended to expand the network of information concerning EMS and identifies the colleges and universities that offer certification and degree programs in the field of EMS.

U.S. Fire Administration, Federal Emergency Management Agency. *EMS Safety--Techniques and Applications*. Report No. FA- 144/April 1994. Emmitsburg, Maryland: USFA/FEMA, 1994.

Comprehensive manual that addresses the hazards faced by EMS providers and describes ways in which emergency responders can and should deal with these risks.

U.S. Fire Administration, Federal Emergency Management Agency. *Guide to Developing and Managing an Emergency Service Infection Control Program*. Report No. FA-112March 1992. Emmitsburg, Maryland: USFA/FEMA, 1992.

Guidebook to assist managers of emergency service organizations in developing, implementing, managing, and evaluating infection control programs.

Vineyard, Sue. *Beyond Banquets, Plaques and Pins: Creative Ways to Recognize Volunteers*. Boulder, Colorado: Volunteer Management Associates, 1988.

Techniques for recognizing and responding to the needs of volunteers.

Zimmerman, Jr., Don. "Rewarding Your Volunteers--A Retirement System Any Department Can Afford," *Firehouse* (March 1993):46(2).

Proposal for a retirement system for volunteer fire departments in which a department's investment in a volunteer's retirement is determined by a activity-based point system. The cash value per point is computed by dividing the funds available for retirement investment by the total points earned by all volunteers.

Volunteerism

Council for Volunteerism. "For the Newcomer: A Brief Look at Volunteer Administration." *Voluntary Action Leadership* (Winter 89-90):23(5).

Concise presentation of important issues in volunteer administration: recruiting, interviewing, orientation, training, supervision, relationships, recognition, firing, and evaluation.

Eden, Dov, and Kinnar, Joseph. "Modeling Galatea: Boosting Self Efficacy to Increase Volunteering." *Journal of Applied Psychology* 76 (1991):770(11).

Report of a study of the Galatea effect, which is a boost in performance caused by raising volunteers' self-expectations.

Hutchinson, David C. "Writing Rules for Volunteers." *Fire Chief* (July 1987):52.

Importance of a clear definition of rules and roles for the professionalization of an organization and the value of maximum volunteer participation in the development of the rules in increasing efficiency.

Krajick, Kevin. "Fighting Fire With . . . What?" *Newsweek* 122 (December 13, 1993):730.

Description of the decline in the number of volunteer firefighters and the problems of volunteer fire companies.

Mabra, Russell. "The American Volunteer: Unpaid Professional." *Firehouse* 17 (March 1992).32(2).

Description of the typical volunteer, the reasons for volunteering, and methods for keeping volunteers involved.

McCurley, Steve. *Volunteer Management Policies.* Boulder, Colorado: Volunteer Management Associates, 1990.

Brief guidelines for conducting volunteer programs: management procedures, recruitment and selection, training and development, supervision and evaluation, and volunteer support and recognition.

McCurley, Steve, and Vineyard, Sue. *101 Ideas for Volunteer Programs.* Downers Grove, Illinois: Heritage Arts Publishing, 1986.

Checklists and suggestions concerning the management of volunteer programs.

Snook, Jack W. "Volunteerism in America: Past, Present and Future." Capstone paper presented at Lewis and Clark College, May 1991.

Overview of the importance of American volunteerism. The concerns and problems currently experienced by organizations using volunteers and the difficulties expected in recruiting, training, and retaining volunteers in the future are identified.

Thompson III, Alexander M., and Bono, Barbara A. "Alienation, Self-actualization, and the Motivation for Volunteer Labor." *Review of Radical Political Economics* 24 (Summer 1992): 115(9).

Exploration of the issue of worker motivation by examining volunteer firefighters and why they engage in this essential activity without pay.

Thompson III, Alexander M., and Bono, Barbara A. "Work Without Wages: The Motivation for Volunteer Firefighters." *The American Journal of Economics and Sociology 52* (July 1993):323(21).

Theoretical and empirical exploration into the motivation of volunteer firefighters.

"Volunteer Program Guide." *Corrections Today* 55 (August 1993) : 66(5).

Guidelines for starting and maintaining a volunteer program, identification of the qualities of a good volunteer, and some suggestions and rules for volunteers.

Wright, Brenda White. "Some Notes on Recruiting and Retaining Minority Volunteers." *Voluntary Action Leadership* (Winter 1992):20.

Fourteen tactics for increasing minority volunteer involvement.

Resources

American Association of Retired Persons (AARP)
1909 K Street, N.W.
Washington, DC 20049
(202) 434-2277

Florida EMS Clearinghouse
1317 Winewood Boulevard
Tallahassee, FL 32399
(904) 487-1911

International Association of Fire Chiefs
EMS Section/Volunteer Section
4025 Fair Ridge Drive, Suite 300
Fairfax, VA 22033
(703) 273-0911

National Association of Emergency Medical Technicians
9140 Ward Parkway
Kansas City, MO 64114
(800) 346-2368

National Emergency Training Center
Learning Resource Center
16825 South Seton Avenue
Emmitsburg, MD 21727
(800) 638-1821 (outside Maryland)
(301) 447-1030 (Maryland only)

National Highway Traffic Safety Administration
Office of Enforcement and Emergency Services
EMS Division
400 Seventh Street, S. W.
Washington, DC 20590
(202) 366-5440

National Volunteer Fire Council
1050 17th Street, N.W., Suite 701A
Washington, DC 20002
(202) 887-5700

Points of Light Foundation
(includes former National Volunteer Center)
1737 H Street, N.W.
Washington, DC 20006
(800) 879-5400

Service Corps of Retired Executives (SCORE)
409 Third Street, S.W.
Washington, DC 20024
(202) 205-6762

U.S. Fire Administration
Fire Technical Programs
16825 S. Seton Avenue
Emmitsburg, MD 21727
(301) 447-1231

www.ingramcontent.com/pod-product-compliance
Lightning Source LLC
Chambersburg PA
CBHW081133170526

45165CB00008B/2653